Cell Planning for Wireless Communciations

For a complete listing of the *Artech House Mobile Communications Library*, turn to the back of this book.

Cell Planning for Wireless Communciations

Manuel F. Cátedra
Jesús Pérez-Arriaga

Artech House
Boston • London

Library of Congress Cataloging-in-Publication Data
Cátedra, Manuel F.
Cell planning for wireless communications / Manuel F. Cátedra and Jesús Pérez-Arriaga.
p. cm. — (Artech House mobile communications library)
Includes bibliographical references and index.
ISBN 0-89006-601-9 (alk. paper)
1. Cellular telphone systems — Planning. 2. Personal communication service systems. I. Pérez-Arriaga, Jesús. II. Title. III. Series.
TK5103.485.C38 1999
621.3845'6—dc21 99-10825
CIP

British Library Cataloguing in Publication Data
Catedra, Manuel F.
Cell planning for wireless communications. – (Artech House mobile communications library)
1. Cellular radio
I. Title II. Perez-Arriaga, Jesus
621.3 ' 84 ' 56

ISBN 0-89006-601-9

Cover design by Lynda Fishbourne

685 Canton Street
Norwood, MA 02062

International Standard Book Number: 621.3 ' 84 ' 56
Library of Congress Catalog Card Number: 99-10825

Contents

Preface

The objective of this book is to provide the reader with details about most of the main tools used in the planning of micro/picocellular systems for personal communication networks. Our approach is based on the extensive use of computer codes that perform reliable analyses of the different aspects that play a significant role in micro/picocellular design: propagation, fading estimation, mutual interference, cell geometry, teletraffic, channel allocation, etc. Some of these aspects are reviewed in Chapter 1.

To a large extent the text has been dedicated to describing how to compute propagation in micro/pico cells. The usage of a combination of deterministic propagation models, topographical/morphological databases and ray-tracing acceleration techniques is presently improving the design of microcell network systems. The most extensive chapters of the book, Chapters 2 and 3, deal with these topics. The idea is to present in an easy but complete way the main algorithms of ray tracing: physical optic (PO), geometrical optic (GO) and the uniform theory of diffraction (UTD), in a manner the reader can understand as well as to develop tools for the analysis of propagation in complex urban or indoor scenarios.

Although the main task of the book is devoted to deterministic approaches, a survey of some empirical propagation models is presented in Chapter 4. These models are easier to apply than the deterministic ones and they do not require a large amount of data, as do the deterministic ones. Quite often the empirical propagation models give enough information, or in any case, complementary information that can be quite useful for the designer.

In the past, most of the parameters of the radio channel had been obtained from experimental data. But now that reliable deterministic models are available, these parameters can also be obtained using data from computer simulations.

In this way, new aspects of channel behavior can be obtained as well as the cost for channel characterization being dramatically reduced if such extensive measurement tests are avoided. Chapter 5 outlines how to obtain some channel parameters from data obtained using a UTD model. The reader can extend the examples presented in Chapter 5 to obtain other channel parameters.

Chapter 6 is dedicated to the new issues in cell architecture and on the channel planning that appear in microcells in relation to macrocells. These new issues are not only governed by the particular characteristics of microcells but also by the new possibilities that arise thanks to the usage of computer tools for the accurate evaluation of propagation losses in complex urban or indoor environments. Among the new issues that are addressed in Chapter 6 we can find: the geometry of the cells, the design of the architecture of the cellular system, the evaluation of the traffic vector and of the compatibility matrices, as well as the usage of computer algorithms for channel allocation in the cell of a system following fixed or dynamic channel assignment strategies.

The book is addressed to researchers and engineers working on wireless communication but it also can be used by students completing graduate courses in radiocommunication systems, mobile communications, antennas and propagation electromagnetic compatibility, etc. The reader is assumed to have basic knowledge of antennas and propagation and of telecommunication systems. Although the book is focused on the design of micro/picocellular systems, most of the approaches and algorithms presented can be employed for macrocellular systems.

The authors would like to acknowledge the invaluable help provided by many people in the development and execution of this book. In the first place, we wish to recognize the contribution of the researchers and students of our team; particularly, Olga Conde, Francisco Saez de Adana, Oscar Gutierrez, Ivan Gonzalez, Javier Cantalapiedra, Miguel Rico, Alberto Gonzalez, Luis Velarde, and Carmen Garcia. The authors are grateful to the many people of Telefónica Moviles: Cayetano Lluch, Eduardo Alonso, Juan Luis Lopez, José Manuel Cousillas, and Pedro Navarro for their support and confidence in our work. Acknowledgment also goes to Roberto Garcia, Cristian Gabetta and Fernando Abalos from Motorola Spain for the support and the databases provided. We also thank Artech House's anonymous reviewers. Finally the authors want to express their gratitude to their linguistic adviser, Mr. Scott Bango, for helping to express their ideas in English.

M. Felipe Cátedra
Jesús Pérez-Arriaga
Alcalá, February 1999

1

Introduction

1.1 Cells, Microcells, and Picocells

Nowadays, the dream that any communication user may have access to any other user "anytime, anywhere" to exchange "any kind" of information appears not to be so far away thanks to the cellular concept in wireless communications [1–6]. Voice, data, video, or a combination of them, generated by a person or a machine, will be sent to any other person or machine using cheap, lightweight portable terminals and sharing intelligent communication systems with an extremely low cost per unit of transported information.

To achieve this huge communication capacity, the deployment of several cellular network levels is foreseen that will cover the populated areas of the Earth with overlapping between them [7–9]. Figure 1.1 illustrates these levels of cellular networking covering an urban residential area. The mobile station (MS) can choose any one of the three links indicated: (a) the urban microcellular low-powered wireless network, (b) the regional cellular network, or (c) the continental or world-based network. The three networks indicated in Figure 1.1 will maintain the MS link in most of the MS locations. When the MS is in an urban area with a very high density of users, it will most probably be linked by a microcellular base station. This link will be the cheapest and probably the only way to satisfy the huge number of users in urban areas, whereas, on the other hand, the propagation environments will probably only allow a link with a close base station (e.g., indoor picocells). However, when the MS suffers signal fading from the picocell network, or when this network is temporarily unable to support the traffic density demand, or when the MS leaves the urban area over which the picocellular network extends, the MS will probably connect with the macrocellular-based network.

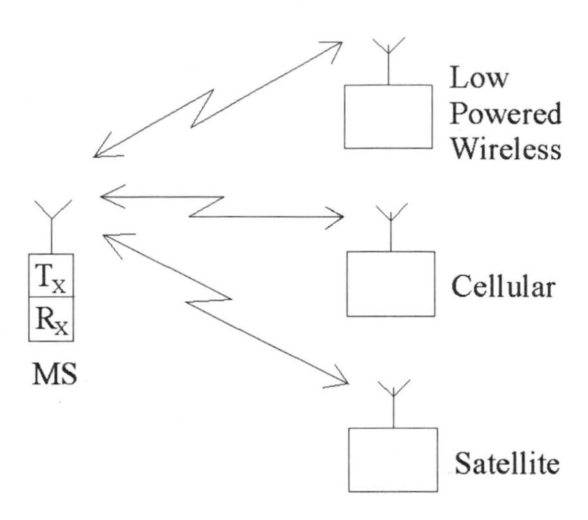

Figure 1.1 A mobile station (MS) will be linked to the world using any one of the three wireless communication networks indicated: satellite, cellular, and microcellular (or picocellular).

Coexisting within the cellular network, the satellite system will improve the MS link when power coverage drops out within the cells or when high-intensity traffic blocks the capacity of the cellular and picocellular nets to attend to new users. Lots of possibilities for satellite systems appear to exist in the future when extensive use of these systems can technically and economically compete with ground-based systems. In any case, due to the special environments of most picocells (indoor, around skyscrapers, etc.), it appears unlikely that the satellite system could eliminate the ground system, based on low-powered wireless equipment.

To cover a large area with a reduced number of channels, presumably with a great number of users, any one of the three systems indicated in Figure 1.1 must employ the cellular concept [10, 11]. Using this concept, the territory covered by a system is split into subareas; in each one of these subareas a cluster of cells is deployed. All the channel resources are used in each cluster. Users in different clusters can share (reuse) the same channel because they are in cells with a large enough distance between them to avoid mutual interference. The calls in a cluster are managed by a switching center (SC). Each SC can manage one or more clusters. The users of this system are linked with the rest of the "world" through the SC, which can connect with a higher node of the mobile network or with the public switched telephone network [12, 13].

One important concern in wireless systems is the mobility of the users. When one user is linked to the system and moves from one cell to another cell, the system should assign a new channel in the arriving cell so that the user does not lose the link. In wireless literature this is called a *handoff* or *handover.* Mobility is always a concern. It is a concern in large cells because the user can be onboard a fast-moving vehicle, whereas in picocells systems a pedestrian can cross several cells in the course of a call.

One of the main goals in the design of a wireless network is to satisfy the traffic demand of all the potential users with a minimum standard of quality in the received signal and with an acceptable percentage of blocked net connections. The quality of the signal is strongly determined by the propagation conditions between the transmitters (desired and interferers) and the receiver. Obviously, fades in the signal strength should be minimized.

The percentage of blocking is usually measured by the grade of service (GOS) [14, 15], which takes two important parameters into account: (1) the unattended new call attempts and (2) the calls in progress dropped when no channel is assigned to a mobile changing of the cell (handoff). Attending the channel requirements for handoffs is usually a priority.

Although the procedure and criteria for designing any one of the three kinds of wireless networks indicated in Figure 1.1 have much in common, the special conditions of the microcellular scenarios impose their own methodology for dealing with these last systems. The design of a microcellular system must consider special aspects of many of the parameters of the network, including these:

- The parameters for the description of the geometry and morphology of the microcell should be more detailed and more accurate than for macrocells.
- The electromagnetic mechanisms that couple the receiver with the transmitter are defined by the direct ray and simple reflected rays in a large percentage of the microcell points.
- The statistical behavior of the field levels and of the other channel parameters. Many times, statistical behavior of the field strength in microcells is described by the Rice distribution [12, 13].
- The aspects of the cell clusters and the carrier-to-interference ratios. Neither the cell geometry nor the cluster configuration have a regular shape that repeats itself along the system. Even more, the concept of the cluster can lose its sense in microcell systems.
- The traffic demand variations in the different cells and the time variations of the traffic are considerably larger in microcells than in macrocells.

These aspects require the application of a new methodology for the analysis and design of picocellular networks, that is, by necessity, different from the well-established approaches currently in use to deal with macrocellular systems.

Because most of the developments throughout the book are applicable for both microcells (in few words, cells of several hundreds of meters of extension) and picocells (cells of tens of meters in size), both terms are used interchangeably in this text.

1.2 Book Structure

This book attempts to present some important aspects of the design of picocellular or microcellular networks, focusing its effort on the practical details more than on theoretical developments.

Chapter 2 describes some of the more important deterministic models used to analyze propagation in picocells. Due to the nature of the picocell scenarios either in urban areas or in indoor spaces, the field coverage is based on a few, prominent coupling mechanisms such as direct rays, reflected rays, and simple diffracted rays, etc., which are caused by the interaction of the radiated field with structures close to the Tx (transmitter) or Rx (receiver) antennas (walls, edges of wedges, furniture, etc.). These coupling mechanisms and their corresponding fields can hardly be described in terms of only two-dimensional (2D) statistical or empirical models. Instead, three-dimensional (3D) deterministic models must be used in combination with accurate geometrical and morphological representations of the picocell scenes. Two deterministic models are reviewed: physical optics and the geometrical theory of diffraction in its uniform formulation (GTD/UTD). Our endeavor focuses on the GTD/UTD model, because from a practical point of view, it is the model that best combines accuracy and computational efficiency. The GTD/UTD model is developed with enough detail that the reader should have enough information to develop his or her own computer tool for the analysis of propagation.

Chapter 3 deals with two very important practical items in computer tools for propagation analysis: the input data of the picocell scene and the management of the ray path computation (ray-tracing). Two formats, Drawing Interchange File (DXF) and Initial Graphic Exchange Specification (IGES), are the most widely used for the 3D representation of the geometry and morphology of picocellular scenes. These formats are able to exhaustively represent all kinds of information concerning indoor or urban scenes, but most of this information is irrelevant for propagation analysis. The propagation prediction tools require only the information relative to the geometric and morphological characteristics of the scene.

Most of Chapter 3 is dedicated to describing ray-tracing algorithms for the computation of propagation at microwave or higher frequencies. We use the term *ray-tracing algorithms* to refer to the procedures that speed up the computations of the ray paths. In a complex scene with hundreds or thousands of facets and edges, rapid computation of the ray path is essential to be able to apply a deterministic model. Using an efficient ray-tracing algorithm, the CPU time and the memory requirements can be reduced by orders of magnitude. The shooting and bouncing method, binary space partitioning, volumetric space partitioning and the angular z-buffer ray-tracing algorithms are presented in Chapter 3. All of these methods have their own pros and cons and, possibly, a good computer tool for propagation will be based on a combination of some of them.

When the wireless network designer studies the interferences between distant picocells or between picocells and macrocells, probably the best way to compute the propagation losses can be to use an empirical or semiempirical method. This is because in this case, the computational resources that a deterministic approach requires are very large and the improvement in accuracy is negligible. Chapter 4 presents a list of the most commonly used empirical or semiempirical approaches for the computation of propagation in macrocells, minicells, and picocells, including indoor scenarios. The chapter development is focused on the practical aspects of the application of these algorithms.

The wireless system or equipment designer requires good knowledge of the parameters and functions that describe the behavior of the propagation channel in the picocells. Assuming that a GTD/UTD model is accurate enough, one could expect that such parameters and functions can be obtained from the model. Chapter 5 presents a detailed discussion showing how to obtain the baseband impulse response of the propagation channel from GTD/UTD data. The chapter is completed with some notes of how to derive first and second statistical parameters of the channel from results of a GTD/UTD model.

Picocell networks have special features that require new procedures for the base station site location and for channel assignment strategies. The special propagation conditions in the picocells usually give geometric shapes for the cells that differ from the classical hexagonal cells. Additionally, because propagation is not described by a regular expression that relates the losses with the distance between the base and the mobile stations, new design rules are needed to select the antenna site and the antenna orientation. Chapter 6 attempts to answer some of these questions, presenting and discussing maps of antenna coverages and cell geometries in urban areas with regularly spaced streets and blocks. From the discussion, some conclusions for cellular design in urban areas can be obtained. As an example of a cell design, an irregular downtown area is presented. Chapter 6 finishes by reviewing some important issues in frequency planning for picocells: the computation of the compatibility matrices and the application of fixed and dynamic channel assignment algorithms.

References

[1] Schneiderman, R., *Wireless Personal Communications. The Future of Talk,* Piscataway, NJ: IEEE Press, 1994.

[2] Pahlavan, K., and A. Levesque, *Wireless Information Network,* New York: John Wiley, 1995.

[3] Feher, K. *Wireless Digital Communications,* London: Prentice Hall, 1995.

[4] Grubb, J. L., "The Traveller's Dream Come True," *IEEE Communications Mag.,* Vol. 29, No. 11, Nov. 1991, pp. 48–51.

[5] Cox, D. C., "Wireless Personal Communications: What Is It?" *IEEE Personal Communications,* Vol. 2, No. 2, April 1995, pp. 20–35.

[6] Padgett, J. E., C. G. Günther, and T. T. Hattori, "Overview of Wireless Personal Communications," *IEEE Communications Mag.,* Vol. 33, No. 1, January 1995, pp. 48–51.

[7] Yeung, K. L., and S. Nanda, "Channel Management in Microcell/Macrocell Cellular Radio Systems," *IEEE Trans. on Vehicular Technology,* Vol. 45, No. 4, November 1996, pp. 601–612.

[8] Mohorcic, M., G. Kandus, E. Del Re, and G. Giambene. "Performance Study of an Integrated Satellite/Terrestrial Mobile Communication System," *Int. J. Satellite Commun.,* Vol. 14, 1996, pp. 413–425.

[9] Steele, R., "Microcellular Radio Communications," Chap. 20 in *The Mobile Communications Handbook,* J. D. Gibson, Ed., Boca Raton, FL: CRC Press and IEEE Press, 1996.

[10] MacDonald, V. H., "Advanced Mobile Phone Service: The Cellular Concept," *Bell Syst. Tech. J.,* Vol. 58, January 1979, pp. 15–41.

[11] Yacoub, M. D., "Cell Design Principles," Chap. 19 in *The Mobile Communications Handbook,* J. D. Gibson, Ed., Boca Raton, FL: CRC Press and IEEE Press, 1996.

[12] Lee, W. C. Y., *Mobile Cellular Telecommunications Systems,* New York: McGraw-Hill, 1990.

[13] Lee, W. C. Y., *Mobile Communications Design Fundamentals,* 2nd ed., New York: John Wiley, 1993.

[14] Feuerstein, M. J., and T. S. Rappaport, *Wireless Personal Communications,* Boston: Kluwer Academic Publishers, 1993.

[15] Steele, R., *Mobile Radio Communication,* London: Pentech Press, 1992.

2

Review of GTD/UTD and Physical Optics Techniques

Deterministic radio propagation prediction in urban and indoor scenarios is a highly complicated electromagnetic problem. The complexity of the scenario (random in certain aspects) makes it impossible to predict radio propagation with a high degree of accuracy. Still, depending on the characteristics of the environments, certain techniques may or may not be suitable. Several techniques have been used in the deterministic and semi-deterministic models: geometrical theory of diffraction (GTD), physical optics (PO), and, not as frequently, rigorous methods such as integral equation (IE) methods or finite-difference time-domain (FDTD).

Rigorous techniques are difficult to use because of the electric size of the environmental obstacles. Such methods require a discretization of the obstacles in elements with dimensions lower than a fraction of the wavelength (typically $\lambda/8$). The number of resulting elements for a typical urban or indoor scene at personal communication networks (PCNs) working frequencies become enormous. Consequently, the memory size requirements and the CPU times generally make these techniques infeasible in conventional 3D environmental models. The rigorous techniques can be used in small picocells [1, 2] or to predict the effects of electrically small objects in the environment. In [3] a simplified 2D method based on an IE method is proposed.

The majority of the deterministic and semi-deterministic propagation models utilize asymptotic high-frequency methods such as GTD, in its uniform theory of diffraction (UTD) version, and PO. In recent years, such techniques have been widely used with good results. This chapter focuses on the application of such techniques in deterministic and semi-deterministic models. The goal of the chapter is not to present a deep and general revision of the GTD and PO techniques; adequate references for that are found in [4–7]. As Chapter 3 shows, the environments are geometrically described using faceted models so many of the GTD and PO expressions are particularized to flat faceted obstacles.

The GTD approach is a ray-based technique. The electromagnetic field is calculated as a sum of the individual contributions associated with the rays. In their propagation, the rays interact with obstacles suffering reflections, diffractions, etc. Sections 2.1 and 2.2 present the laws that govern field propagation considering the antennas as field sources and relating their radiation patterns to the propagated field. Section 2.3 deals with application of the image theory to ray reflections. Practical expressions of the reflected field are derived as a function of the transmitter antenna radiation pattern and the material properties of the reflecting surfaces. The edge-diffraction phenomenon (including slope-diffraction) is analyzed in Section 2.4 where the GTD/UTD expressions are particularized to straight edges, which are the main type of edges in urban and indoor scenarios. Section 2.5 presents a geometric optics (GO) solution for the transmitted field calculation. Section 2.6 deals with multiple effects, that is, combinations of more than two reflections, diffractions, and transmissions. Sections 2.7 and 2.8 address the application of the GTD approach in deterministic and semi-deterministic propagation models. The PO approach has also been widely used in propagation prediction both in semi-deterministic and deterministic models. Sections 2.9, 2.10, and 2.11 deal with the application of PO-based techniques in such models.

2.1 GTD/UTD Approach

The electric field $\vec{E}_T$ created at the observation point O by a source located at S will be approximated by the series:

$$\vec{E}_T = \sum_{i=1}^{N} \vec{E}_i \tag{2.1}$$

where $\vec{E}_i$ represents the electric field due to each one of the N ray paths that connects point S with O: direct ray, reflected rays, diffracted rays, transmitted rays, reflected-diffracted rays, double-reflected rays, etc. Figure 2.1 illustrates some of the ray paths connecting the source with the observation point in a simple model of an urban scenario: direct ray (dir), reflected ray (ref), and edge-diffracted ray (dif).

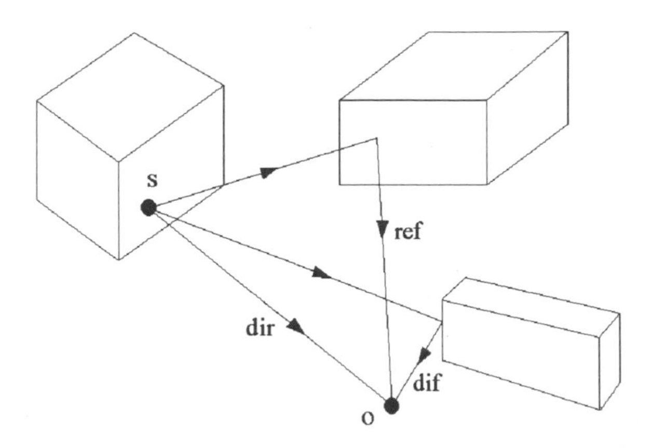

Figure 2.1 Example of three rays reaching the observation point O.

Each one of the terms in the series of (2.1) can be computed using the ray tube formulation of GTD/UTD. The electric field associated with a propagating ray is given by

$$\vec{E}(s) = \vec{E}(0)\sqrt{\frac{\rho_1\rho_2}{(\rho_1 + s)(\rho_2 + s)}}\exp(-j\beta s) \tag{2.2}$$

where $\vec{E}(s)$ is the electric field at a point at a distant s to the reference point ($s = 0$), $\vec{E}(0)$ is the electric field at the reference point, and ρ_1, ρ_2 are the principal radii of curvature of the wavefront associated with the ray at the reference point, which depends on the caustic lines (see Figure 2.2).

In (2.2), β is the free-space wave number, which is given by

$$\beta = \frac{2\pi}{\lambda} \tag{2.3}$$

where λ is the wavelength. Two particularly interesting cases are the spherical and the cylindrical wave ray tubes:

- Spherical wave ray tube. This case occurs when both caustic lines degenerate to the same point, so the wavefronts are spherical. Consequently, $\rho_1 = \rho_2 = \rho_0$ and the electric field at s becomes

$$\vec{E}(s) = \vec{E}(0)\frac{\rho_0}{\rho_0 + s}\exp(-j\beta s) \tag{2.4}$$

When the caustic point is taken as the reference point, (2.4) can be expressed as follows [4]:

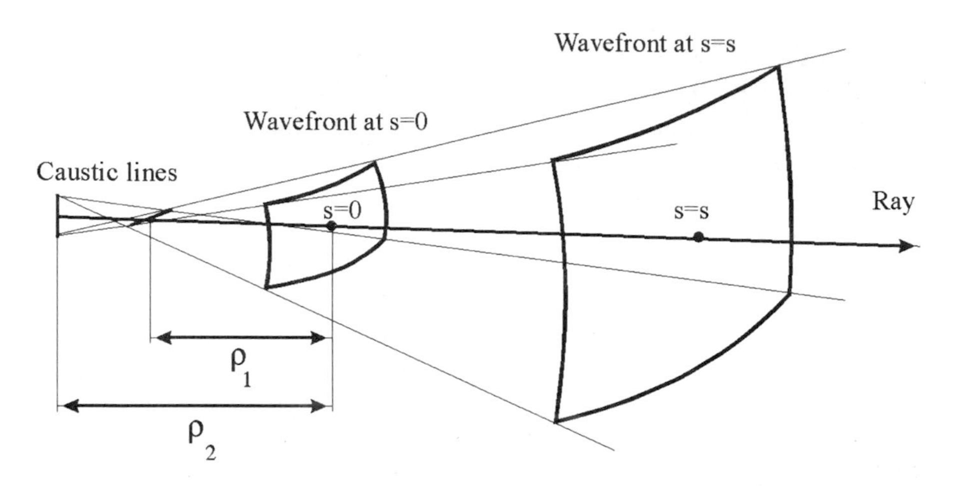

Figure 2.2 Ray tube propagation.

$$\vec{E}(s) = \vec{E}_0 \frac{\exp(-j\beta s)}{s} \tag{2.5}$$

where $\vec{E}_0$ is a vector that can be viewed as an excitation factor for the spherical ray tube.

- Cylindrical wave ray tube. This case occurs when one of the caustic lines is at infinity so, $\rho_1 = \infty$ and, consequently, the wavefronts are cylindrical. In this case the electric field at s becomes

$$\vec{E}(s) = \vec{E}(0)\sqrt{\frac{\rho_2}{\rho_2 + s}}\exp(-j\beta s) \tag{2.6}$$

When the reference point is at the caustic line, (2.6) can be written as follows [4]:

$$\vec{E}(s) = \vec{E}_0 \frac{\exp(-j\beta s)}{\sqrt{s}} \tag{2.7}$$

In this equation, $\vec{E}_0$ can be viewed as an excitation factor for the cylindrical ray tube.

When a ray reaches an obstacle, a reflected ray arises (see Figure 2.1). The GO laws of reflection give the characteristics of the reflected ray: the direction of propagation, the principal radii of curvature of the reflected wavefront, and the field at the new reference point, which is the reflecting point. Thus, the reflected ray propagates as (2.2) states. When a ray reaches an edge (see Figure 2.1), an infinite number of diffracted rays arises. Similarly, as before, GTD/UTD laws provide the characteristics of the emerging diffracted rays [8].

The conventional GTD/UTD are applicable when the incident field is a ray-optical field [4], that is, when it can be expressed as in (2.2). In the mobile communication context, the preceding condition is satisfied except in certain special cases, as pointed out later in Sections 2.4 and 2.6.

Assuming that in the geometric model the facets are flat and therefore all the edges are straight, the wavefronts are simpler than in Figure 2.2: spherical wavefronts for the direct and reflected rays and nearly cylindrical wavefronts for the case of diffracted rays when the observation point is far away from the edge.

2.2 Direct Field

The field contribution at point O due to the direct ray is given by

$$\vec{E}(O) = \vec{E}_t(\theta,\phi)\frac{\exp(-j\beta r)}{r} \tag{2.8}$$

where

$$\vec{E}_t(\theta,\phi) = \sqrt{\frac{\eta P_r G}{2\pi}}\vec{E}_o(\theta,\phi) \tag{2.9}$$

and where η is the free-space impedance, P_r is the power radiated by the transmitter, G is the gain of the transmitter antenna, and $\vec{E}_o(\theta,\phi)$ is the normalized radiation pattern of the transmitter antenna.

The spherical coordinates (r,θ,ϕ) of the observation point O refer to a coordinate system associated with the antenna. Usually the geometrical parameters of the facets refer to a different system (geometry coordinate system). A transformation of coordinates between the geometry and the antenna systems needs to be made for the field computation. To do this, the utilization of the director cosines is recommended (see Appendix 2A).

Expressions (2.8) and (2.9) can be used when the observation point is in the far field of the antenna, which is typical of mobile communications. However, this can be untrue if an array or an electrically large antenna is employed. In these two cases, several phase centers must be considered. The field contribution due to an antenna with N phase centers is given by

$$\vec{E}(O) = \sum_{i=1}^{N} \sqrt{\frac{\eta P_{ri} G_i}{2\pi}} \vec{E}_o(\theta_i, \phi_i) \frac{\exp(-j\beta r_i)}{r_i} \tag{2.10}$$

2.3 Reflected Field

To obtain the reflected field when a ray reaches a surface, one can use the general expressions of GO when they are applicable (e.g., all the facets are large enough in terms of wavelengths, the transmitter antenna is far away from the surface, and the radii of curvature of the surfaces at the reflection point are electrically large). These expressions are valid for both the curved and flat surfaces of a microcell. Assuming that the environmental obstacles have no curved surfaces and that the GO is applicable, a most efficient way to compute the field contribution of the reflected rays is to apply image theory and the Fresnel reflection coefficients instead of the general GO expressions [4–6]. The image theory can be derived from the GO laws when the surfaces are plane so, in such a case, both methods will provide the same results. Hereafter, we concentrate on the application of the image theory to the reflection problem.

Figure 2.3 shows a reflection case. The facet in the figure belongs to a body whose permittivity, conductance, and permeability are defined by parameters ϵ, σ, and μ, respectively. The antenna radiation pattern $\vec{E}_o(\theta, \phi)$, is defined relative to the antenna axes shown in the figure. The phase center of the antenna (S) is assumed to be located in the origin of its coordinate system.

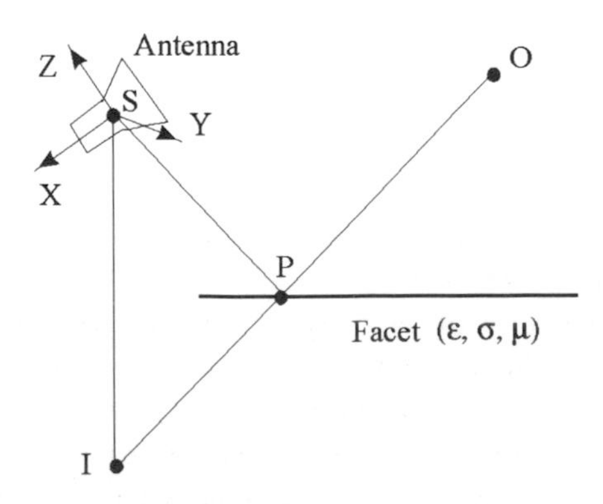

Figure 2.3 Transmitter antenna in front of a flat facet and definition of the image point and the reflection point.

Following Fermat's principle, the reflected ray path SPO indicated in Figure 2.3 can be obtained by minimizing the length of this path. In other words, the incident ray is reflected in the specular direction (this is also known as Snell's law). Observe that point P can also be obtained as the intersection point between the straight line IO and the facet, I, as being the image of the source point S. Points S, P, O, and I are in the same plane, which is known as the incident plane. The incident plane is always perpendicular to the facet plane.

To efficiently compute the reflection fields for a large number of observation points, we will use the so-called facet-fixed axes (X_f, Y_f, Z_f) of Figure 2.4. From these axes we can obtain the image-fixed axes (X_i, Y_i, Z_i) as defined in Figure 2.5.

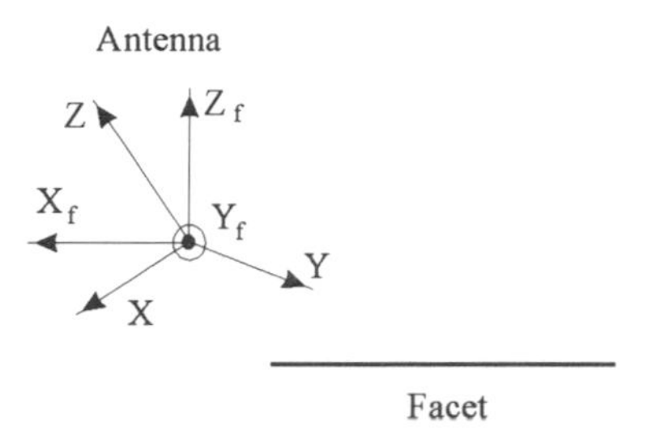

Figure 2.4 The facet-fixed axes (X_f, Y_f, Z_f) share the origin with the antenna-fixed axes but the Z_f axis is perpendicular to the facet. The facet-fixed axes can be defined in a preprocessing step by making Z_f parallel to the normal vector to the facet, X_f parallel to one of the sides of the facet, and Y_f perpendicular to the other two axes.

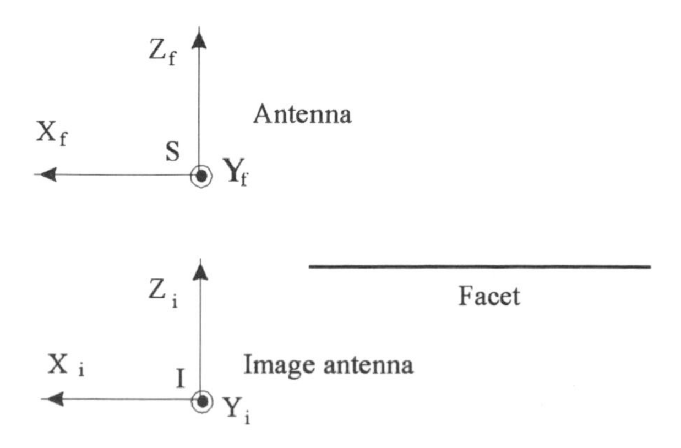

Figure 2.5 The image axes are defined by a translation of the facet-fixed axes. The center of the image axes is located in the image I of the antenna phase center S.

Using the facet-fixed axes, it is a straightforward process to expand the incident field in the so-called parallel and perpendicular (to the plane of incidence) polarization vectors because now the spherical unit vectors $\hat{r}_f, \hat{\theta}_f, \hat{\phi}_f$, are parallel to the so-called ray-fixed axes [4]. The vertical and perpendicular field components are also called hard and soft, respectively. In other references, they also appear as vertical and horizontal components. It is obvious that $\hat{r}_f$ is parallel to the incident wave direction, $\hat{\phi}_f$ is parallel to the soft polarization vector $\hat{u}_s^i$, and $\hat{\theta}_f$ is parallel to the hard polarization vector $\hat{u}_h^i$ (see Figure 2.6).

In a similar manner, one can expand the reflected field in the soft (perpendicular) and hard (parallel) components. In Figure 2.6 these vectors are denoted as $\hat{u}_s^r$ and $\hat{u}_h^r$, respectively. These vectors are directly related to the spherical unit vectors $\hat{r}_I, \hat{\theta}_I, \hat{\phi}_I$ of the image-fixed systems: $\hat{u}_s^r$ and $\hat{u}_h^r$ are parallel to $\hat{\phi}_I$ and $\hat{\theta}_I$, respectively and $\hat{r}_I$ is parallel to the reflected ray direction (see Figure 2.6). The soft components of the incident and reflected fields coincide $\hat{u}_s^i = \hat{u}_s^r$.

Using the facet-fixed and the image-fixed coordinate system, the incident field at P is given by

$$\begin{aligned}\vec{E}_i(\mathrm{P}) &= [E_{\theta f}(\theta_f,\phi_f)\hat{\theta}_f + E_{\phi f}(\theta_f,\phi_f)\phi_f]\frac{\exp(-j\beta r_f)}{r_f} \\ &= [-E_{\theta f}(\theta_f,\phi_f)\hat{u}_h^i - E_{\phi f}(\theta_f,\phi_f)\hat{u}_s^i]\frac{\exp(-j\beta r_f)}{r_f}\end{aligned} \tag{2.11}$$

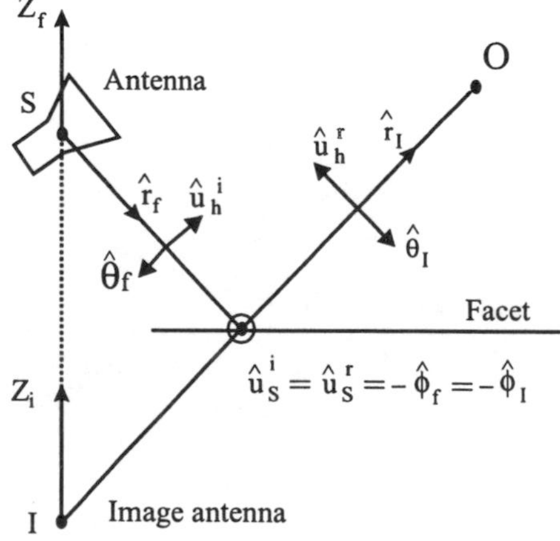

Figure 2.6 Soft (perpendicular) and hard (vertical) polarization vectors for the incident and reflected rays.

where (r_f, θ_f, ϕ_f) are the spherical coordinates of P in the facet-fixed coordinate system and $E_{\theta f}$ and $E_{\phi f}$ are the field components of the transmitter antenna radiation pattern relative to the facet-fixed axes.

The reflected field at P can be obtained from the incident field by using the Fresnel reflection matrix $\overline{\Gamma}$:

$$\vec{E}_r = \overline{\Gamma}\vec{E}_i \tag{2.12}$$

When $\vec{E}_i$ and $\vec{E}_r$ are resolved into their soft and hard components, (2.12) can be written as follows:

$$\begin{bmatrix} E_s^r \\ E_h^r \end{bmatrix} = \begin{bmatrix} \Gamma_s & 0 \\ 0 & \Gamma_h \end{bmatrix} \begin{bmatrix} E_s^i \\ E_h^i \end{bmatrix} \tag{2.13}$$

where Γ_h and Γ_s are the so-called soft and hard reflection Fresnel coefficients, respectively. Then, the following GO approximation for the reflected field at the observation point O can be obtained:

$$\vec{E}_r(\mathrm{O}) = [\Gamma_h(\theta_I)E_{\theta f}(\pi - \theta_I, \phi_I)\hat{\theta}_I + \Gamma_s(\theta_I)E_{\phi f}(\pi - \theta_I, \phi_I)\hat{\phi}_I]\frac{\exp(-j\beta r_I)}{r_I} \tag{2.14}$$

where (r_I, θ_I, ϕ_I) are the spherical coordinates of the observation point in the image-fixed system.

The Fresnel coefficients are given by

$$\Gamma_h(\theta) = \frac{\epsilon_r \cos(\theta) - \sqrt{\epsilon_r - \sin^2(\theta)}}{\epsilon_r \cos(\theta) + \sqrt{\epsilon_r - \sin^2(\theta)}} \tag{2.15}$$

$$\Gamma_s(\theta) = \frac{\cos(\theta) - \sqrt{\epsilon_r - \sin^2(\theta)}}{\cos(\theta) + \sqrt{\epsilon_r - \sin^2(\theta)}} \tag{2.16}$$

where θ is the angle of incidence formed by the incident ray and the normal vector to the facet. Notice that θ coincides with θ_I, and ϵ_r is the complex relative dielectric constant:

$$\epsilon_r = \frac{\epsilon - j\dfrac{\sigma}{\omega}}{\epsilon_0} \tag{2.17}$$

Lists of the electromagnetic properties of building materials can be found in [9].

Note that the reflected field at the observation point can be computed considering an equivalent antenna at the image point I, with the following radiated pattern referred to the image-fixed system:

$$\vec{E}_I(r_I,\theta_I,\phi_I) = \vec{E}_{tI}(\theta_I,\phi_I)\frac{\exp(-j\beta r_I)}{r_I} \tag{2.18}$$

where

$$\vec{E}_{tI}(\theta_I,\phi_I) = \Gamma_h(\theta_I)E_{\theta f}(\pi - \theta_I,\phi_I)\hat{\theta}_I + \Gamma_s(\theta_I)E_{\phi f}(\pi - \theta_I,\phi_I)\hat{\phi}_I \tag{2.19}$$

Thus, it can be stated that the reflected field is equivalent to the direct field of an antenna whose associated coordinated system is the image-fixed system, and its pattern is given by (2.18). This means that for a second-reflection problem, the same procedure can be applied as for the first-reflection case, but now we start from the equivalent image antenna.

One of the assumptions required in the application of the Fresnel coefficients is that the reflecting surfaces must be smooth. In the case of such surfaces, the incident waves are reflected only in the specular direction. In the case of rough surfaces, not all the energy is concentrated in the specular direction, and a portion of the incident wave energy is reflected at other angles. Therefore, when the Fresnel coefficients are applied to rough surfaces, one must expect an overestimation of the reflected field amplitude. Following the Raleigh criterion, a surface is considered rough if the maximum and minimum surface variations (h) satisfy

$$h > \frac{\lambda}{8\cos\theta} \tag{2.20}$$

with λ being the incident field wavelength and θ the angle of incidence.

Some authors [10] propose a correction in the Fresnel coefficients that will account for the surface roughness. This correction consists of multiplying both coefficients by a factor $(1 + \rho)/2$, where ρ is the so-called scattering loss factor given by

$$\rho = \exp\left[-8\left(\frac{\pi\sigma_h\cos\theta}{\lambda}\right)^2\right] \tag{2.21}$$

where σ_h is the standard deviation of the surface height. Equation (2.20) is derived assuming that the surface heights are Gaussian distributed. A more sophisticated (and accurate) expression for the scattering loss factor can be found in [10, 11]. The scattering loss factor, when applied in a GTD/UTD model, corrects the amplitude of the specularly reflected field but does not account for the field reflected out of the specular direction (diffuse reflection).

2.4 Edge-Diffracted Field

The field diffracted by an edge is given by [8]

$$\vec{E}^{d}(s) = \vec{E}^{i}(Q_d)\overline{D}\sqrt{\frac{\rho_e^i}{s(\rho_e^i + s)}}e^{-j\beta s} \tag{2.22}$$

where $\vec{E}^{i}(Q_d)$ is the incident field at the diffraction point Q_d, $\overline{D}$ is the diffraction coefficients matrix, and s is the distance between the observation point (O) and the diffraction point. To apply (2.22), the incident field must refer to the edge-fixed coordinate system associated with the incident plane. This is the plane that contains the edge and the incident ray (see Figure 2.7). The unit vectors of the edge-fixed coordinate system are: the vector in the

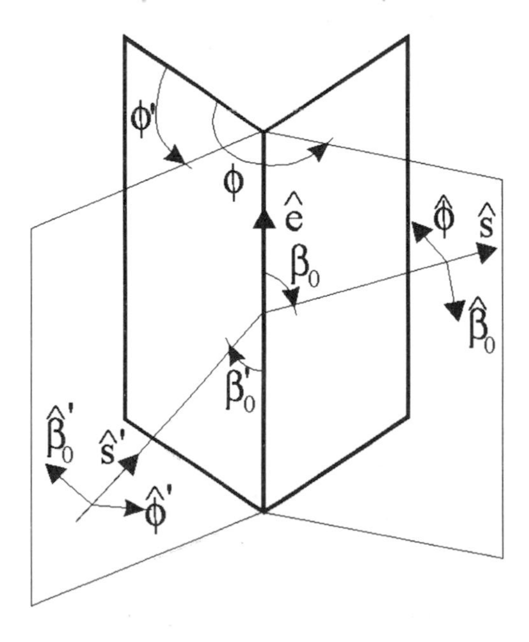

Figure 2.7 Diffraction by a straight wedge: definition of the edge-fixed axes.

direction of the incident ray $\hat{s}'$, the vector perpendicular to the incident plane $\hat{\phi}'$, and the vector parallel to the incident plane $\hat{\beta}_0'$ (see Figures 2.7 and 2.8). These last two vectors can be obtained with the following expressions:

$$\hat{\phi}' = -\frac{\hat{e} \times \hat{s}'}{|\hat{e} \times \hat{s}'|} \qquad \hat{\beta}_0' = \hat{\phi}' \times \hat{s}' \tag{2.23}$$

where $\hat{e}$ is a vector along the edge.

The diffracted rays propagate from the diffraction point in the directions ($\hat{s}$), which satisfy

$$\hat{s}' \cdot \hat{e} = \hat{s} \cdot \hat{e} \tag{2.24}$$

The infinite rays that fulfill this equation form a cone whose vertex is the diffraction point (see Figure 2.8). This is the so-called "Keller's cone."

The term ρ_e^i in (2.22) is the radius of curvature of the incident wave in the plane of incidence. When the incident wavefront is spherical, it coincides with the distance between Q_d and the transmitter antenna (s').

Equation (2.22) provides the diffracted field referring to the edge-fixed coordinate system associated with the diffraction plane (see Figure 2.7). This is the plane that contains the edge and the diffracted ray. The unit vectors of the associated coordinate system are: $\hat{s}$, vector $\hat{\phi}$ perpendicular to the diffraction plane, and the vector $\hat{\beta}_0$ parallel to such plane (see Figures 2.7 and 2.8). These last two vectors can be calculated as follows:

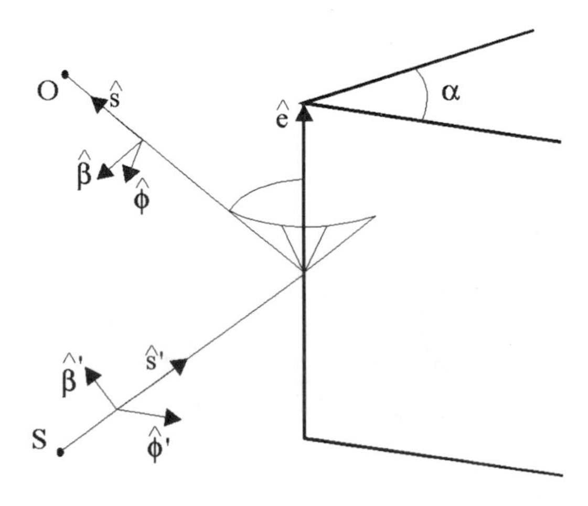

Figure 2.8 Definition of the geometric parameters involved in the diffraction by a straight edge.

$$\hat{\phi} = \frac{\hat{e} \times \hat{s}}{|\hat{e} \times \hat{s}|} \qquad \hat{\beta}_0 = \hat{\phi} \times \hat{s} \tag{2.25}$$

Field components in $\hat{\phi}'$ and $\hat{\phi}$ are perpendicular to the edge. They are the so-called hard components. Field components following $\hat{\beta}_0{}'$, $\hat{\beta}_0$ are the parallel components, also called the soft components.

When the field is resolved into the soft and hard components, the dyadic form of the diffraction coefficients can be expressed as:

$$\overline{D} = \begin{bmatrix} -D_s & 0 \\ 0 & -D_h \end{bmatrix} \tag{2.26}$$

where D_s and D_h are the soft and hard diffraction coefficients, respectively, given by

$$D_{s,h}(L,\phi,\phi',\beta_0,n) = D_1 + D_2 + \Gamma_{s,h}(D_3 + D_4) \tag{2.27}$$

where

$$D_1 = \frac{-e^{-j\pi/4}}{2n\sqrt{2\pi k_0}\,\sin\beta_0}\cot\left[\frac{\pi + (\phi - \phi')}{2n}\right] F[\beta\, La^{+}(\phi - \phi')] \tag{2.28}$$

$$D_2 = \frac{-e^{-j\pi/4}}{2n\sqrt{2\pi k_0}\,\sin\beta_0}\cot\left[\frac{\pi - (\phi - \phi')}{2n}\right] F[\beta\, La^{-}(\phi - \phi')] \tag{2.29}$$

$$D_3 = \frac{-e^{-j\pi/4}}{2n\sqrt{2\pi k_0}\,\sin\beta_0}\cot\left[\frac{\pi + (\phi + \phi')}{2n}\right] F[\beta\, La^{+}(\phi + \phi')] \tag{2.30}$$

$$D_4 = \frac{-e^{-j\pi/4}}{2n\sqrt{2\pi k_0}\,\sin\beta_0}\cot\left[\frac{\pi - (\phi + \phi')}{2n}\right] F[\beta\, La^{-}(\phi + \phi')] \tag{2.31}$$

The angles $\beta_0{}'$, β_0, ϕ', and ϕ are shown in Figure 2.7; β is the wave number and n is a parameter related to the interior wedge angle α (see Figure 2.8) by

$$n = \frac{2\pi - \alpha}{\pi} \tag{2.32}$$

where α is in radians. $F[x]$ is the Fresnel transition function, which is given by

$$F[x] = 2j\sqrt{x}e^{jx}\int_{\sqrt{x}}^{\infty}\exp(-j\tau^2)\,d\tau \tag{2.33}$$

For practical purposes, Appendix 2B shows asymptotic expressions of the Fresnel transition function obtained from [4].

In (2.27)–(2.31), L is the distance parameter given by

$$L = \frac{s(\rho_e^i + s)\rho_1^i\rho_2^i}{\rho_e^i(\rho_1^i + s)(\rho_2^i + s)}\sin^2\beta_0 \tag{2.34}$$

where $\rho_{1,2}^i$ are the radii of curvature of the incident wave front at Q_d. For a spherical incident wavefront $\rho_e^i = \rho_1^i = \rho_2^i = s'$, and the distance parameter can be written as follows:

$$L = \frac{ss'}{s + s'}\sin^2\beta_0 \tag{2.35}$$

The function $a^{\pm}(\delta^{\pm})$ is given by

$$a^{\pm}(\delta^{\pm}) = 2\cos^2\left(\frac{2n\pi N^{\pm} - \delta^{\pm}}{2}\right) \tag{2.36}$$

where $\delta^{\pm} = \phi \pm \phi'$ and $N^{\pm}$ are integer numbers that most closely satisfy the following equations:

$$2\pi nN^{+} - (\delta^{\pm}) = \pi \tag{2.37}$$

$$2\pi nN^{-} - (\delta^{\pm}) = -\pi \tag{2.38}$$

Note that $\Gamma_{s,h}$ represents the Fresnel reflection coefficients of the surfaces of the wedge at the edge.

When the edge-diffracted field is added to the direct and reflected fields, the diffraction coefficients ensure continuity in the field at the shadow boundaries (see Figure 2.9). Surrounding the shadow boundaries are the so-called "transition" regions. The points of the transition regions fulfill $kLa^{\pm} < 2\pi$.

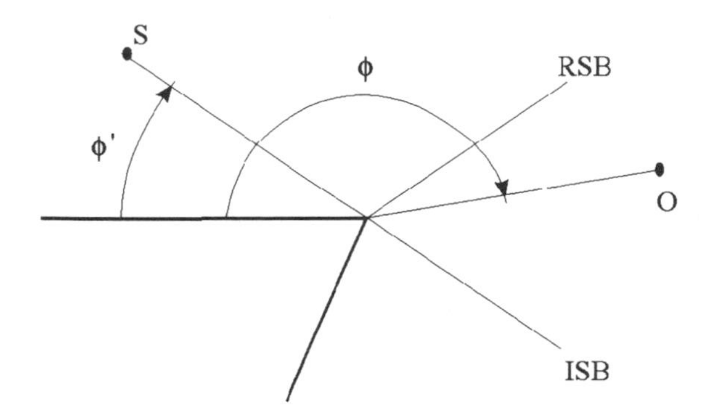

Figure 2.9 Reflection shadow boundary (RSB) and incident shadow boundary (ISB).

Outside the transition regions, the Fresnel transition functions of (2.27)–(2.30) can be approximated [8] by $F(kLa^{\pm}) \approx 1$ so, away from the point of diffraction, the diffracted field (2.22) can be considered a ray-optical field. Otherwise, for observation points inside the transition regions, the resulting diffracted field does not show ray-optical behavior.

Equation (2.22) predicts a null field when the incident field at the diffraction point is zero. To obtain a better prediction in such cases, a second-order contribution must be taken into account. This is known as the slope-diffraction contribution [7] because it is proportional not to the incident field but rather to the slope of the incident field at the diffraction point:

$$\vec{E}^s(s) = \frac{\partial \vec{E}^i(Q_d)}{\partial n}\bar{d}\sqrt{\frac{\rho_e^i}{s(\rho_e^i + s)}}e^{-j\beta s} \tag{2.39}$$

where

$$\bar{d} = \begin{bmatrix} -d_s & 0 \\ 0 & -d_h \end{bmatrix} \tag{2.40}$$

$$d_{s,h} = \frac{1}{j\beta \sin\beta_0}\frac{\partial D_{s,h}}{\partial \phi'} \tag{2.41}$$

$$\begin{aligned}\frac{\partial D_{s,h}}{\partial \phi'} = \frac{-e^{-j\pi/4}}{4n^2\sqrt{2\pi\beta}\sin\beta_0}\Bigg[\Bigg\{&\csc^2\left(\frac{\pi + (\phi - \phi')}{2n}\right)F_s[\beta\, La^+(\phi - \phi')] \\ &- \csc^2\left(\frac{\pi - (\phi - \phi')}{2n}\right)F_s[\beta\, La^-(\phi - \phi')]\Bigg\} \\ &- \Gamma_{s,h}\Bigg\{\csc^2\left(\frac{\pi + (\phi + \phi')}{2n}\right)F_s[\beta\, La^+(\phi + \phi')] \\ &- \csc^2\left(\frac{\pi - (\phi + \phi')}{2n}\right)F_s[\beta\, La^-(\phi + \phi')]\Bigg\}\Bigg]\end{aligned} \tag{2.42}$$

being

$$F_s(x) = 2jx[1 - F(x)] \tag{2.43}$$

The operator $\partial/\partial n$ means the directional derivative at the diffraction point in the direction perpendicular to the plane of incidence, that is, $\hat{\phi}'$. Therefore, it can be expressed as

$$\frac{\partial}{\partial n} = \frac{1}{s'} \frac{\partial}{\partial \phi'} \tag{2.44}$$

2.5 Transmitted Field

When a ray reaches the interface between two media, a portion of the energy is reflected toward the first medium (reflected ray) and a part of the energy is refracted (transmitted) toward the second medium (transmitted or refracted ray). Section 2.3 dealt with the GO approach for the reflected field. The incident and reflected fields were related through the reflection coefficients, which were particularized to the case where the first medium is the free space. In a similar fashion, transmission coefficients can be used to relate the fields associated with the incident and refracted field [5]. The direction of the refracted rays is given by Snell's law of diffraction:

$$\beta_0 \sin\theta_i = \beta \sin\theta_r \tag{2.45}$$

where θ_i is the incident angle formed by the incidence direction and the normal vector at the interface (pointing toward the first medium) in the refraction point, θ_r is the refraction angle formed by the refracted ray and the normal vector at the interface (pointing toward the second medium), and β_0 and β are the wave numbers in the first and second media, respectively (see Figure 2.10).

Moreover, Snell's law states that the incident ray, the normal vector to the interface at the refraction point, and the refracted ray are in the same plane (plane of incidence). This fact, along with (2.45) determines the direction of the transmitted ray.

In the context of outdoor propagation, the influence of the transmitted rays is very low. When a ray is transmitted through an external building wall, it is rapidly attenuated due to the subsequent obstacles that it encounters after the transmission. Therefore, the field of the rays that come back to the outdoor medium are negligible. On the other hand, in the indoor propagation phenome-

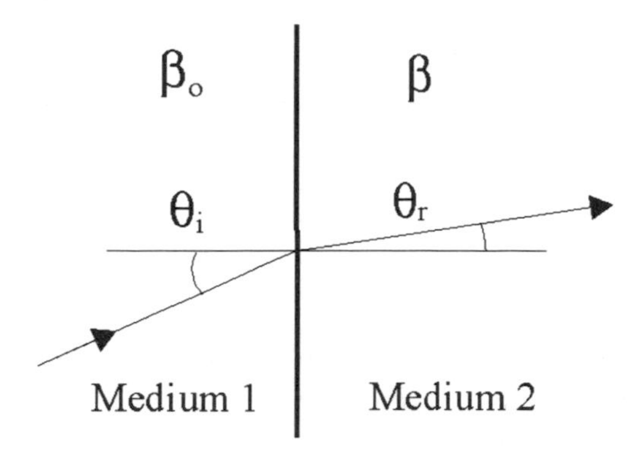

Figure 2.10 Refraction of a ray in the interface between two media.

non, the rays transmitted through the walls play an important role so they must be taken into account in the indoor propagation models. Many authors propose transmission coefficients obtained from measurements with typical building walls [12]. These empirical coefficients are easy to include in propagation models and in many cases they provide good results. But they do not take into account parameters such as the incidence angle, wall width, etc.

The building walls have a finite width such that when a ray is transmitted through a wall, it suffers two transmissions: first, from the exterior medium to the wall medium and, secondly, from the wall medium to the exterior medium. Similarly, the reflection phenomenon in a wall of finite width is different than that in an interface between two media. Hence, to be coherent, when a model considers the wall width in the transmission effect, the reflections must also be treated considering the finite wall width. In the majority of the outdoor models, the reflections in building walls are treated assuming the wall width to be infinite (reflection coefficients of Section 2.3). But, in indoor models, when the transmission simulation considers the finite width of the walls, the reflections must also take into account the real walls.

Approximate wall transmission coefficients given in terms of the reflection coefficients can be found in [1]. In this section, "deterministic" transmission and reflection coefficients for walls of finite width are presented as functions of the electrical properties of the wall and the incident ray. They have been taken from [13, 14] under the following assumptions:

- The wall medium is homogeneous and isotropic.
- The two interfaces are locally plane at the transmission points. This is a common assumption in the context of the GO approach.

These assumptions are not always realistic in an indoor environment but the results obtained (see Section 2.7) reveal this transmission model to be very accurate.

The transmitted field $\vec{E}_i$ through a wall of finite width is

$$\vec{E}_t = \overline{T}\vec{E}_i \tag{2.46}$$

where $\vec{E}_i$ is the incident field at the first interface of the wall and $\overline{T}$ is the transmission matrix. When the incident and transmitted fields are resolved into their soft and hard components, the transmission matrix is given by

$$\overline{T} = \begin{bmatrix} T_s & 0 \\ 0 & T_h \end{bmatrix} \tag{2.47}$$

with

$$T_h = \frac{4\epsilon_r\sqrt{\epsilon_r - \sin^2\theta_i}\cos\theta_i \exp\left[jd\beta_0\left(\sqrt{\epsilon_r - \sin^2\theta_i} + \cos\theta_i\right)\right]}{\exp\left(2jd\beta_0\sqrt{\epsilon_r - \sin^2\theta_i}\right)\left(\epsilon_r\cos\theta_i + \sqrt{\epsilon_r - \sin^2\theta_i}\right)^2 - \left(\sqrt{\epsilon_r - \sin^2\theta_i} - \epsilon_r\cos\theta_i\right)^2} \tag{2.48}$$

$$T_s = \frac{4\sqrt{\epsilon_r - \sin^2\theta_i}\cos\theta_i \exp\left[jd\beta_0\left(\sqrt{\epsilon_r - \sin^2\theta_i} + \cos\theta_i\right)\right]}{\exp\left(2jd\beta_0\sqrt{\epsilon_r - \sin^2\theta_i}\right)\left(\sqrt{\epsilon_r - \sin^2\theta_i} + \cos\theta_i\right)^2 - \left(\cos\theta_i - \sqrt{\epsilon_r - \sin^2\theta_i}\right)^2} \tag{2.49}$$

where θ_i is the incident wave at the first interface of the wall (see Figure 2.11), θ_r is the angle of refraction at the first medium (this can be obtained with (2.44)), d is the wall width, β_0 is the wave number in the free-space, ϵ_r is the

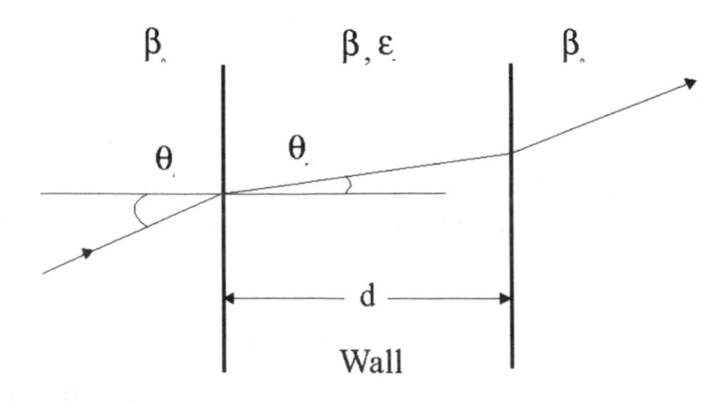

Figure 2.11 Transmission through a wall.

relative permittivity of the wall medium, and β is the wave number of the wall medium:

$$\beta = \frac{2\pi}{\lambda}\sqrt{\epsilon_r} \tag{2.50}$$

where λ is the wavelength in the free space.

The reflected field $\vec{E}_r$ in a wall of finite width can be obtained using (2.12) and (2.13). But now the reflection coefficients are given by

$$\Gamma_s = \frac{(1-\epsilon_r)\left[\exp\left(2jd\beta_0\sqrt{\epsilon_r - \sin^2\theta_i}\right) - 1\right]}{\exp\left(2jd\beta_0\sqrt{\epsilon_r - \sin^2\theta_i}\right)\left(\cos\theta_i + \sqrt{\epsilon_r - \sin^2\theta_i}\right)^2 - \left(\cos\theta_i - \sqrt{\epsilon_r - \sin^2\theta_i}\right)^2} \tag{2.51}$$

$$\Gamma_h = \frac{\left(\epsilon_r - \sin^2\theta_i - \epsilon_r^2\cos^2\theta_i\right)\left[\exp\left(2jd\beta_0\sqrt{\epsilon_r - \sin^2\theta_i}\right) - 1\right]}{\exp\left(2jd\beta_0\sqrt{\epsilon_r - \sin^2\theta_i}\right)\left(\sqrt{\epsilon_r - \sin^2\theta_i} + \epsilon_r\cos\theta_i\right)^2 - \left(\sqrt{\epsilon_r - \sin^2\theta_i} - \epsilon_r\cos\theta_i\right)^2} \tag{2.52}$$

2.6 Multiple Effects

In complex environments such as urban and indoor scenarios, multiple effects (e.g., multiple reflections, multiple diffractions, multiple transmissions, and multiple combinations of reflections and/or diffractions and/or transmissions) must be included to give an accurate estimation of the field at the observation points. As an example, Figure 2.12 shows examples of the double effects involving reflections and/or diffractions. In microcells and picocells, fields representing two or three interactions are sufficient enough to provide a good

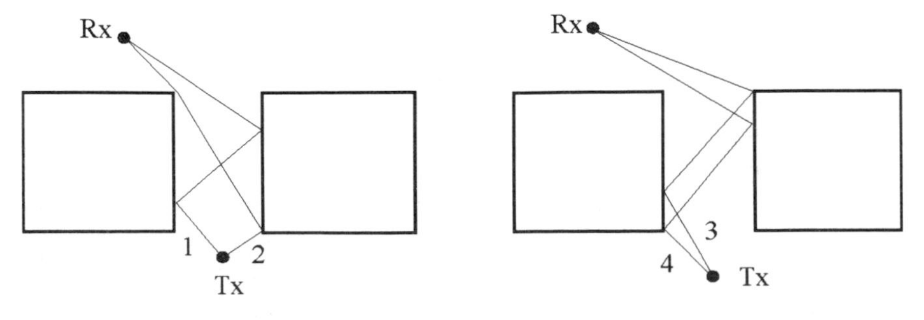

Figure 2.12 Examples of double reflected ray (1), double diffracted ray (2), reflected-diffracted ray (3), and diffracted-reflected ray (4).

answer (except in certain cases such as large indoor corridors). But in large cells, sometimes it is necessary to consider high-order field contributions.

Knowing the reflection, diffraction, or transmission points, a multiple-effect contribution can easily be obtained in a step-by-step manner similar to the analysis carried out for the single reflection, single diffraction, and single transmission effects discussed in previous sections. Chapter 3 presents efficient ray-tracing techniques to determine the reflections, diffraction, and transmission points in multiple effects.

Problems can arise when a ray diffracted by an edge in a transition region acts as an incident ray in a following interaction. As stated in Section 2.4, this is not a ray-optical field, so the conventional GTD/UTD can produce erroneous results. However, in a multiple wedge diffraction when only one transition region is involved, the conventional UTD including slope diffraction yields good results [4, 7]. For the case of double wedge diffraction, Schneider and Luebbers [15] have extended the conventional coefficients to solve this problem. When no transition regions are involved, these coefficients coincide with the product of the conventional diffraction coefficients.

2.7 Validation of GTD/UTD Models

The GTD/UTD model has been widely validated as a technique for deterministic propagation prediction both in indoor [16–20], outdoor [21–30], and rural [27, 31–34] environments. The references reveal the GTD/UTD to be a valid method in very different environments when an adequate description of the environment is available. In this section, additional validations in three different types of scenarios are presented:

1. Urban environment of Manhattan;
2. Urban environment of Madrid;
3. Indoor environment of an office building.

The Manhattan scenario analyzed is an example of an urban environment having buildings of uniform height that are located regularly (see Figure 2.13). The measurements have been taken from [25]. This scenario has also been analyzed in [26] where the authors compare measurements with simulation values obtained from a GTD/UTD-based model. This model is a quasi-3D one in the sense that it does not consider the diffractions from the rooftops. The model accounts for all possible reflections and diffractions in the main street, side streets, and parallel streets. Good accordance between measurements and simulations can be observed.

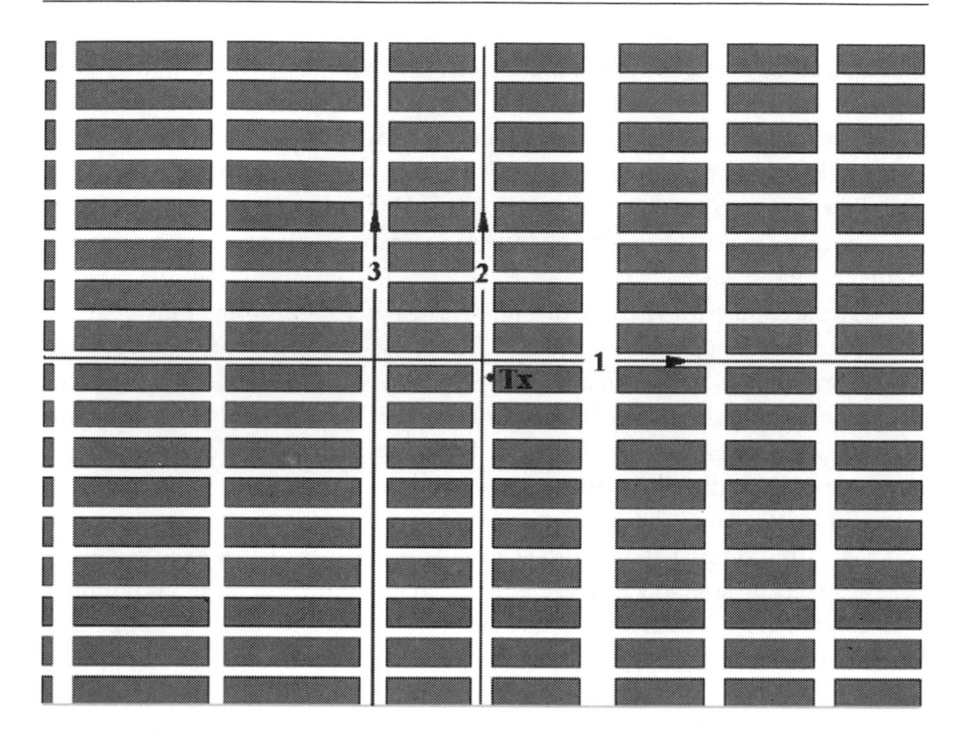

Figure 2.13 Plan view of the Manhattan scenario analyzed. The three lines in the streets indicate the paths analyzed.

In the present simulation, a complete 3D GTD-based code, called FASPRO [28, 29], has been used. The mechanisms considered are first-order effects, second-order effects (except double diffractions), and third-order effects, which involve a ground reflection and do not involve more than one diffraction. Figure 2.13 shows the three paths where the path loss comparison of measurement simulations has been done: Path 1 is along 51st Street, path 2 is along Lexington Avenue, and path 3 is along Third Avenue. The figure also shows the transmitter (Tx) antenna location at Lexington Avenue. In the simulations, the observation points are separated by 2m. The environment contains 136 buildings that have been modeled with 680 facets plus 1 facet for the ground. The number of resulting edges is 1,888. All the buildings are assumed to be of equal height: 25m. The electric properties for all the surfaces, including the ground, have been assumed to be $\epsilon_r = 15$, $\mu_r = 1$, and $\sigma = 7$ S/m. This value of σ is very high compared with the usual conductivity of ground and building walls. For the building walls, the authors have chosen this value from [26]. In [26] there is no information about the ground material so, here, the authors have chosen the same values that were used for the walls. Anyway, for grazing incidences the conductivity values are not important (see (2.15)–(2.17)), so one cannot expect significant changes in the predictions with the exact ground conductivity value.

The Tx antenna is omnidirectional and located at 10m of height. The receiver (Rx) antenna is located at 2m of height. The frequency was 900 MHz.

The first path (path 1 in Figure 2.13) is 1200m long and is perpendicular to the antenna street. The path loss values obtained are compared with measurements in Figure 2.14.

The second path (path 2 in Figure 2.13) is 1300m long. This is in the same street as the antenna so, this is a LOS (line-of-sight) case. The obtained results can be seen in Figure 2.15.

The third path (path 3 in Figure 2.13) is also 1300m long and runs parallel to the antenna street. Figure 2.16 shows the comparison between the measured and predicted path loss.

Table 2.1 shows the mean error predictions (predictions-measurements) and the corresponding standard deviations for the three paths analyzed.

The second scenario analyzed is depicted in Figure 2.17. This corresponds to the commercial center of Madrid (Spain). The results relative to this case have been taken from [29]. Unlike the Manhattan scenario, it is quite irregular, with buildings of different shapes and heights. Figure 2.18 shows a top view map of the environment. The area analyzed is $1050 \times 1450\text{m}^2$ in size and

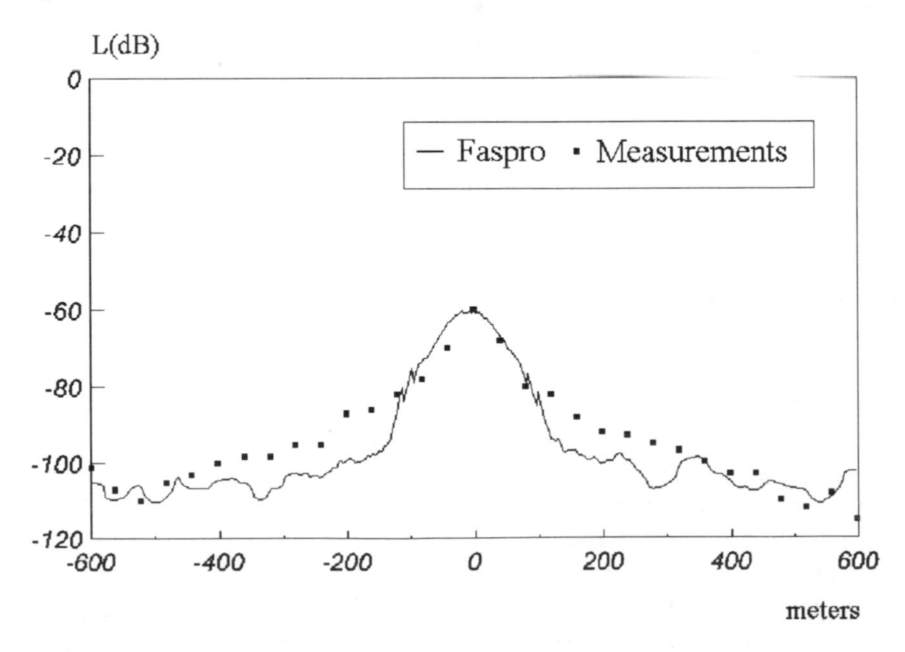

Figure 2.14 Path loss comparison between measurements and computations for the first path of the Manhattan scenario (path 1 in Figure 2.13).

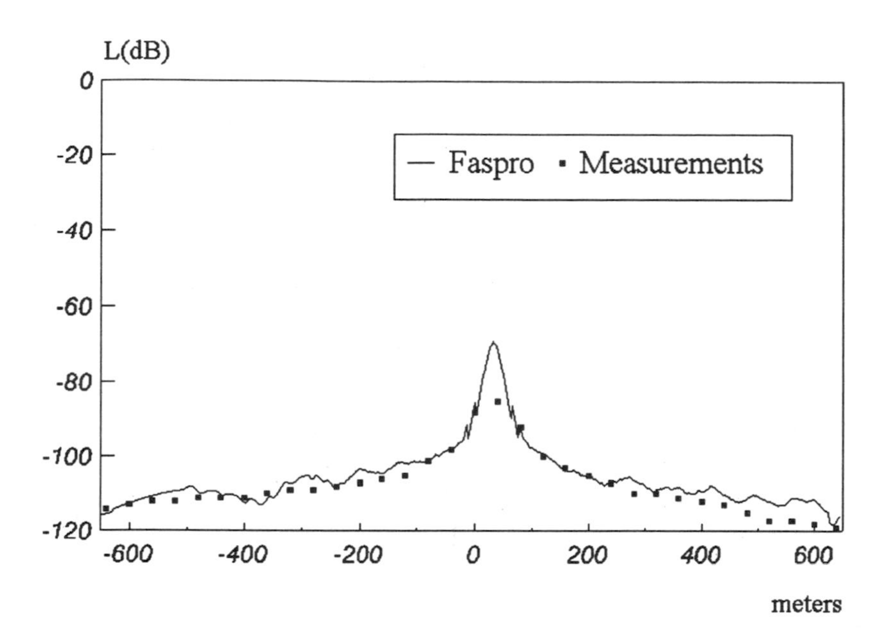

Figure 2.15 Path loss comparison between measurements and simulation for the second path of the Manhattan scenario (path 2 in Figure 2.13).

contains about 100 buildings. The scenario was modeled using about 700 surfaces and 1200 edges with a precision of 1m. No data were available with respect to the buildings and ground material, so in the simulations all of the buildings and the ground were assumed to be made of concrete.

The measurements were made along three paths, indicated as 1, 2, and 3 in Figure 2.18. To avoid the effect of fast fading in the results, for both measurements and computations, the field at each point was averaged with the 10 nearest points in the path. Various samples per wavelength were taken in all the cases, in terms of both measurements and computations. The frequency was 945 MHz.

The Tx antenna was located at a height of 7m in the position indicated in Figure 2.18. The radiated power was 14 dBm and the radiation pattern for the E-plane cut can be seen in Figure 2.19. For the H-plane, the curve is very similar. The receiver antenna was a short dipole always located 1.5 meters high.

The first path, represented as 1 in Figure 2.18, corresponds to a LOS situation. The path length is about 1000m. Figure 2.20 shows the comparison between the measurements and computations considering the direct and the ground reflected rays, that is, the "two-ray model." Both curves are approximately in accordance except for points far away from the transmitter, where

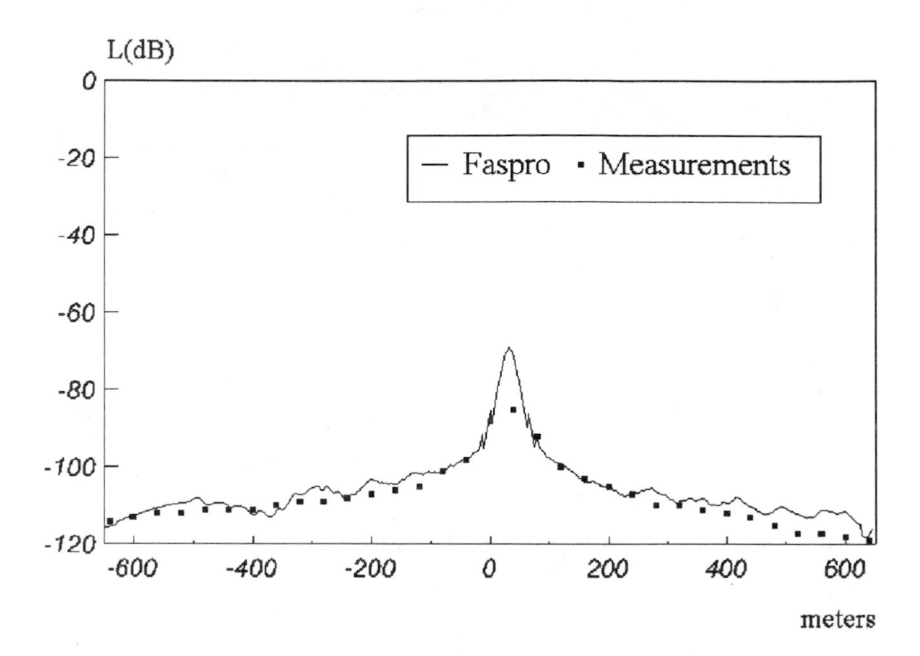

Figure 2.16 Path loss comparison between measurements and simulations for the third path of the Manhattan scenario (path 3 in Figure 2.13).

the field level is low and where other ray mechanisms become important. Figure 2.21 shows the curve obtained considering all the simple mechanisms and the following second-order mechanisms: ground reflection-diffraction, diffraction-ground reflection, and double reflection with one reflection off the ground. The influence of these second-order effects is negligible in points close to the Tx antenna and modifies the simulation values in points far away from the antenna. Finally, Figure 2.22 shows the computation curve considering all the effects already mentioned plus the following double effects: double-reflected rays, diffracted-reflected rays, and reflected-diffracted rays. Triple effects are also included when one of the effects in the triad is a reflection off the ground.

Table 2.1
Mean Error Predictions and Standard Deviations for the Three Paths in Manhattan

	Mean Error	Standard Deviation
Path 1	–3.1	5.6
Path 2	–4.4	3.7
Path 3	2.2	3.1

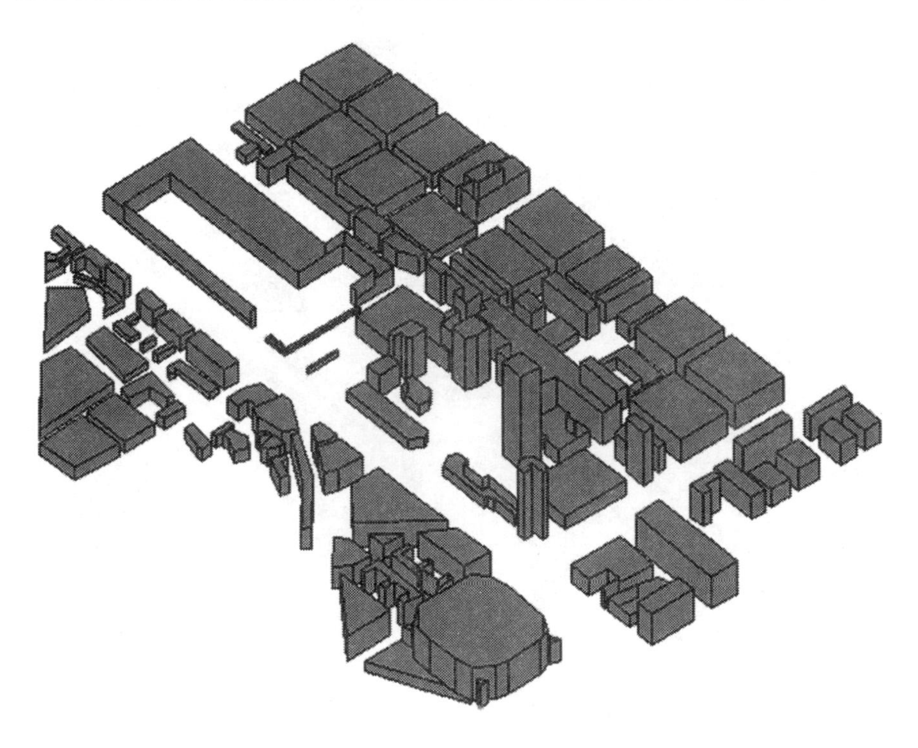

Figure 2.17 Three-dimensional view of the Madrid scenario.

No double diffraction is included because its field level has been found negligible in comparison with the rest of the effects. Once again, the influence of the new effects is practically negligible near the Tx antenna but they improve the prediction in points far away from the Tx antenna.

In the simulations just discussed, the double-diffraction mechanism has not been included because its field level has been found negligible in comparison with the rest of the effects. The discrepancies that arose between measurements and computations near the maximum of the field level can be due to errors in the antenna axis orientation or in the radiation pattern considered.

The second path analyzed (2 in Figure 2.18) is 50m long. This is a non-line-of-sight (NLOS) path where the double and triple effects become more important. Figure 2.23 depicts the computation curve when the effects of Figure 2.22 have been considered.

The third path (path 3 in Figure 2.18) also represents a NLOS case (see Figure 2.24). But in this case the observation points are further away than in the previous path and are severely shadowed with respect to the Tx antenna. The path is 200m long. The mechanisms considered are the same as in the above case.

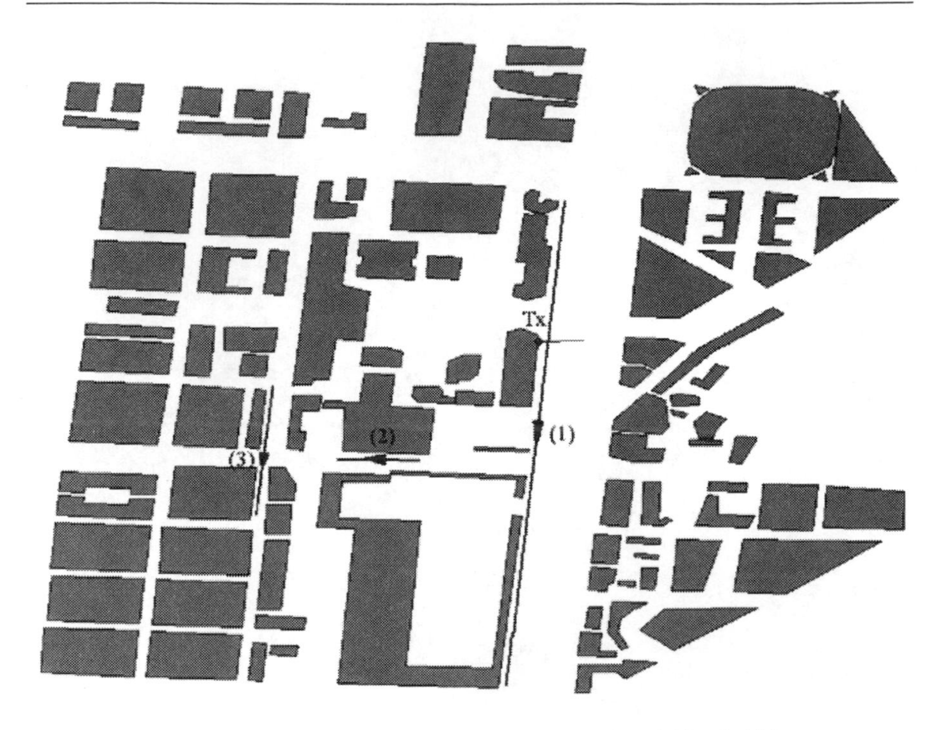

Figure 2.18 Two-dimensional view of the urban scene analyzed. The bold lines represent the paths where the measurements and the simulations were carried out. The transmitter antenna was mounted on a building wall at a height of 7m.

Table 2.2 shows the mean error predictions (predictions-measurements) and the corresponding standard deviations for the three paths analyzed.

To validate the GTD in indoor scenarios, a set of path loss measurements has been made by Telefónica Móviles (a Spanish telephone company). They were compared with a GTD-based simulation code called FASPRI developed by the authors [35]. The path loss measurements were made in the second story of an office building at 900 MHz. The Tx and the Rx antennas were located on the same floor. Only this floor has been considered in the simulations. The receiver power measurements were made along several straight paths for two different locations of the Tx antenna. The Tx antenna is vertically polarized and its radiation (E-plane and H-plane) is depicted in Figure 2.25. Its gain is 7 dBi and the transmitter power was 23 dBm. The Rx antenna was a short dipole always located at 0.6 meters in height.

Figures 2.26 and 2.27 show the geometric model of the scenario with its main dimensions in meters. It contains 175 facets. The height of the ceiling is 2.9m (distance between floors). In the simulation all the facets have the same electric properties ($\epsilon_r = 4.44$, $\sigma = 0.08$ S/m, and $\mu_r = 1.0$) and they

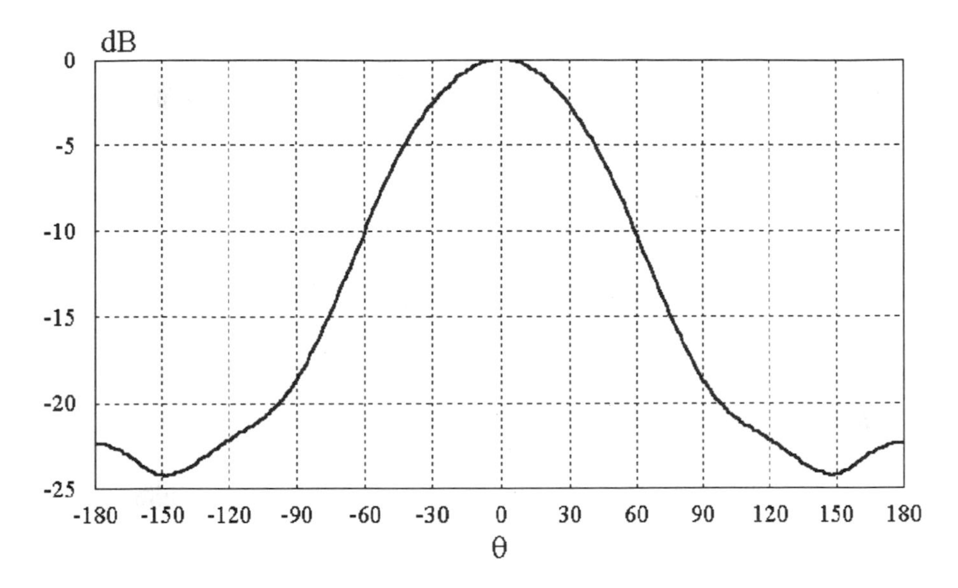

Figure 2.19 E-plane cut of the transmitter antenna radiation pattern. In the H-plane the curve is quite similar.

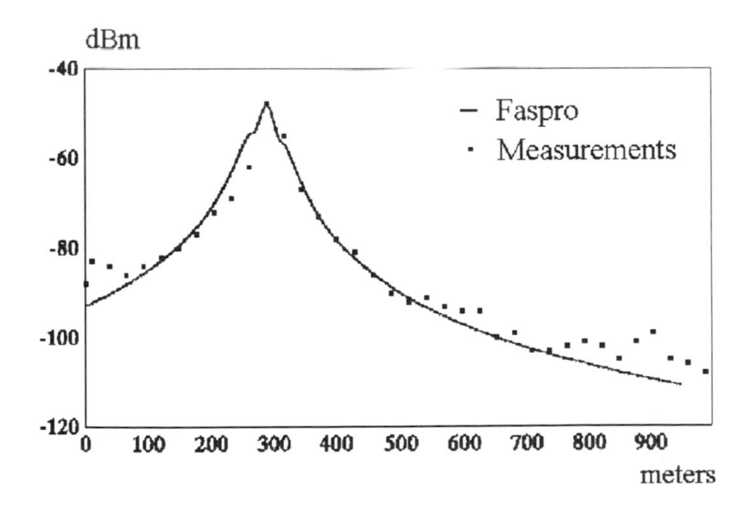

Figure 2.20 Comparison between computations and measurements. Only the direct field and the ground reflected field have been considered in the simulation. (Taken from [29].)

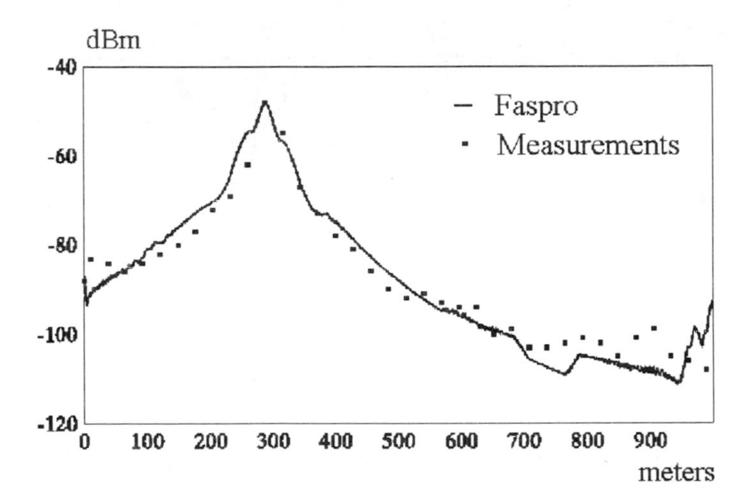

Figure 2.21 Comparison between measurement and simulations considering all of the simple effects and the following second-order effects: ground reflection-diffraction, diffraction-ground reflection, and double reflection with one reflection off the ground. (Taken from [29].)

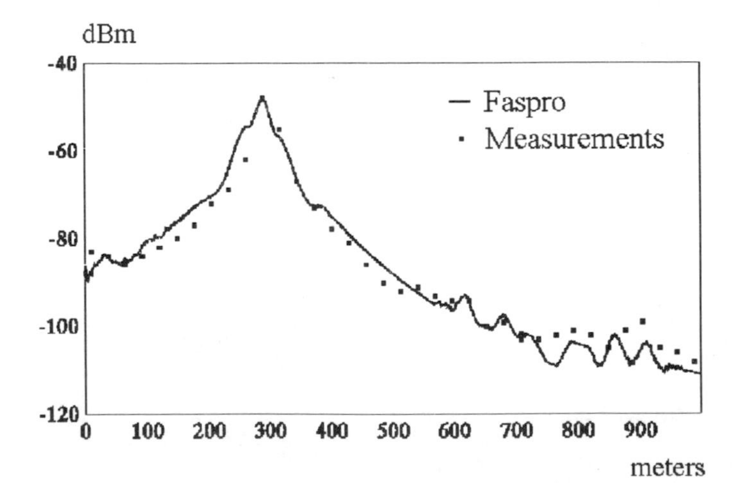

Figure 2.22 Comparison of measurements and simulation taking into account all of the mechanisms considered in Figure 2.21 plus the following double effects: double-reflected rays, diffracted-reflected rays, and reflected-diffracted rays. Triple effects are also included when one of the effects in the triad is a reflection off the ground. (Taken from [29].)

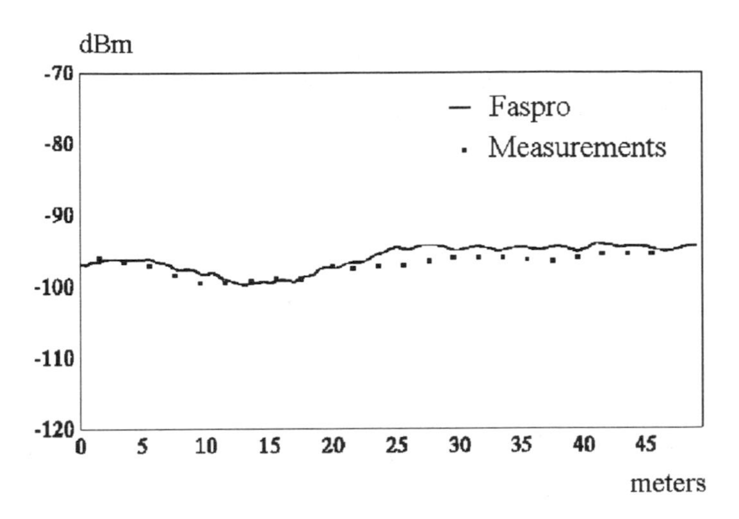

Figure 2.23 Comparison between computations along path 2. All the simple, double, and triple effects defined in Figure 2.22 have been considered. (Taken from [29].)

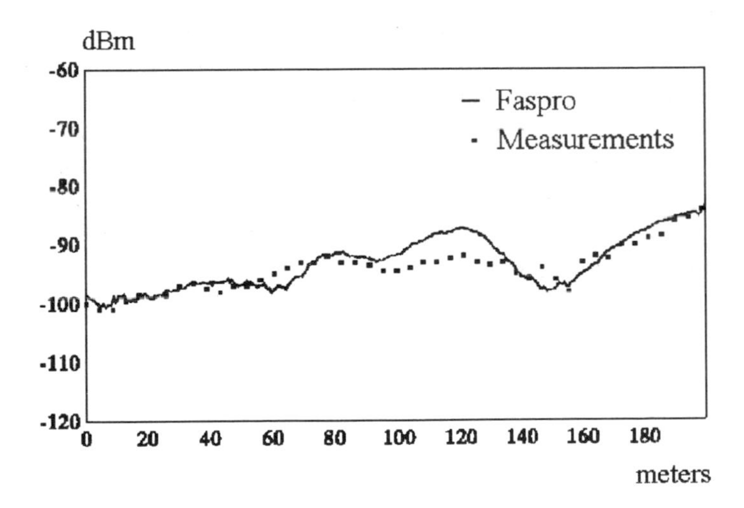

Figure 2.24 Computation and measurements for path 3. The simulation considers all the effects considered in Figure 2.22. (Taken from [29].)

were considered smooth. The width of the walls was assumed to be 0.1m. The mechanisms considered were:

- Simple effects;
- Double effects (all the possible combinations involving reflections and/or diffractions and/or transmissions);

Table 2.2
Mean Prediction Errors and Standard Deviations for the Three Paths in Madrid
(Simulations take effects considered in Figure 2.22 into account)

	Mean Error	Standard Deviation
Path 1	−0.39	3.88
Path 2	1.24	1.12
Path 3	0.96	2.35

- Triple reflections;
- Third-order effects involving one, two, or three transmissions; and
- Fourth-order effects involving two, three, or four transmissions.

In the first three cases the Tx antenna is supported by a column and located at (5.67, 13.85, 1.7), where the coordinates are in meters. It is orientated along the Y axis (the direction of maximum radiation is parallel to the Y axis).

In the first case, the measurements were made in 71 points in both directions: left to right (path 1a in Figure 2.27) and right to left (path 1b in Figure 2.27). Sometimes, as in this case, the measurements along a path are carried out in both directions to avoid errors due to the fast fading. The path length was 30.0m. Figure 2.28 shows the comparison between measurements and the simulation for path 1a. Figure 2.29 compares measurements and simulation values for path 1b.

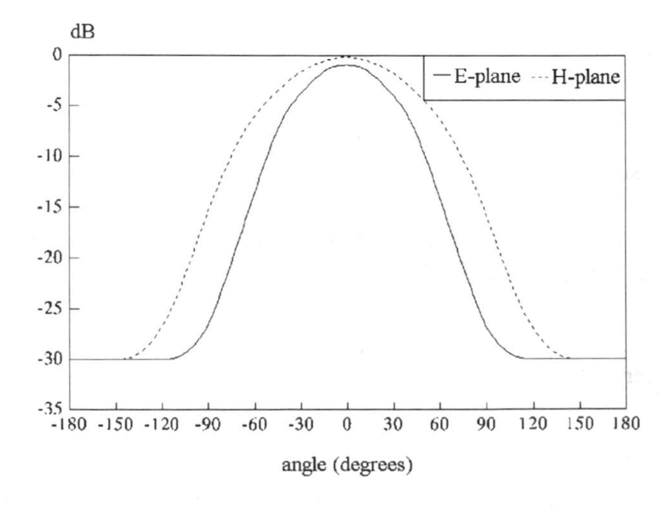

Figure 2.25 Radiation pattern of the transmitter antenna: E-plane cut and H-plane cut.

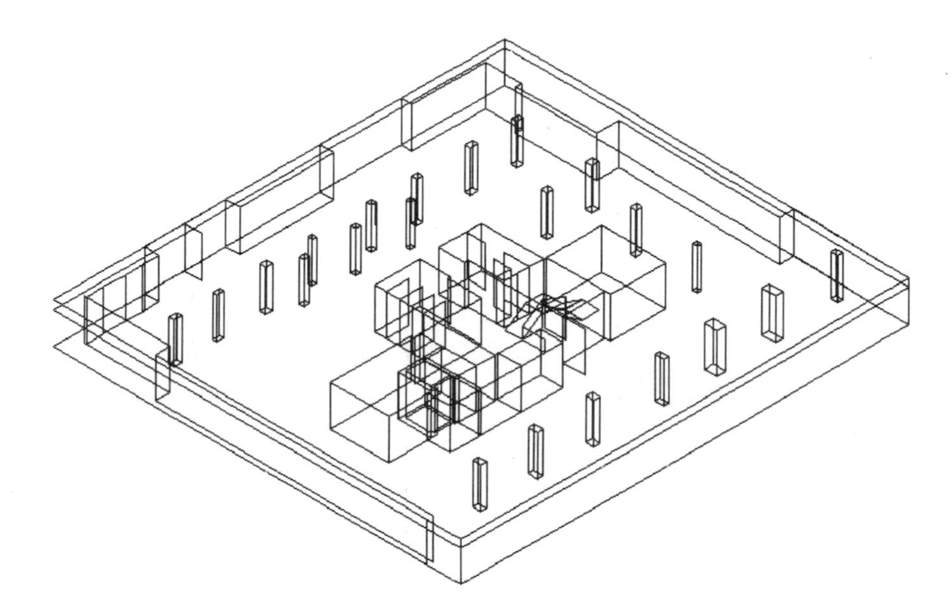

Figure 2.26 Three-dimensional view of the indoor scenario.

The second path (path 2 in Figure 2.27) is 28.3m long. The measurements were made at 61 points. The comparison between simulation and measurements for this LOS path are presented in Figure 2.30.

The third path (path 3 in Figure 2.27) corresponds to a NLOS situation. The path length is 23.0m. The measurements were made at 48 points. Figure 2.31 represents the values of the measurements and simulation.

Other measurements were made with the same Tx antenna located at (21.0, 21.5, 2.85) and oriented along the −Z axis downward (the coordinates are in meters). Figure 2.32 shows the antenna location and the two paths analyzed.

Path 4 is 25.7m long and the measurements were made at 43 points. Figure 2.33 shows the measurement and the simulation values for this case.

Path 5 corresponds to a NLOS situation. The measurements were made at 42 points and the path length is 25.3m. Measurements and simulation are presented in Figure 2.34.

Table 2.3 shows the mean error predictions (predictions-measurements) and the corresponding standard deviations for the six cases analyzed.

In general, the accuracy of the GTD/UTD-based deterministic models are limited by:

- The inherent error of the GTD/UTD approach;
- The approximation in the treatment of the reflection, diffraction, and transmission in dielectric materials;

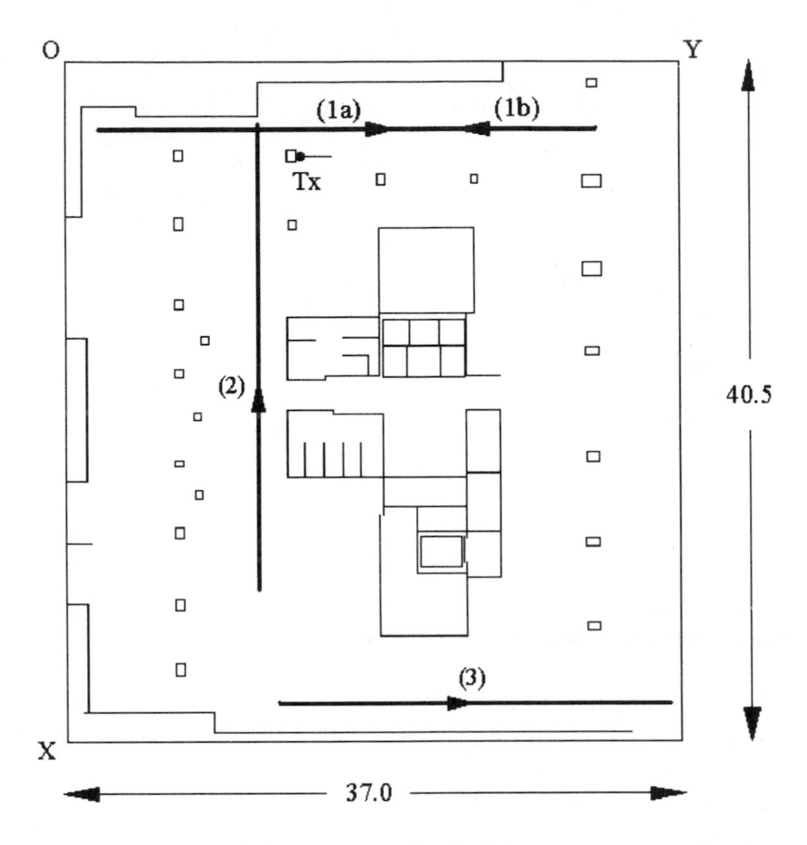

Figure 2.27 Top view of the indoor scenario. The bold lines represent the first three paths analyzed. The Tx antenna was located at (5.67, 13.85, 1.7) and it was oriented parallel to the Y axis (all the coordinates are in meters).

- The inability to consider an infinite number of mechanisms;
- The assumption that the surfaces reflect only in specular directions, that is, they do not give diffuse reflections; and
- The accuracy in the morphological and topographical description of the environment: electric properties, the approximation of curved surfaces by plane facets, objects not considered in the model, etc.

2.8 Application of GTD/UTD on Semi-Deterministic Models

One of the earliest semi-deterministic models that used the GTD approach was proposed by Ikegami et al., [36]. It assumes a NLOS situation where the transmitter antenna is above the rooftops of the buildings close to the receiver

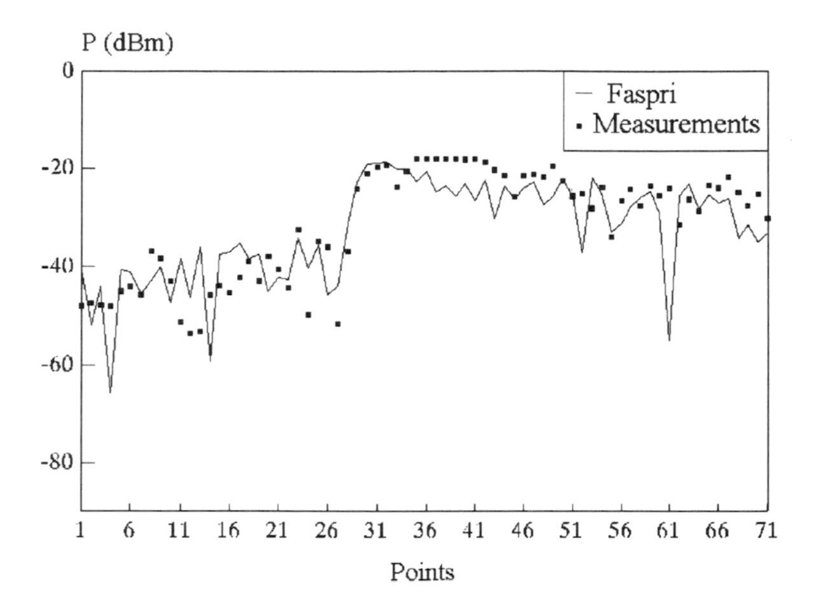

Figure 2.28 Comparison between measurements and simulation for LOS path 1a of Figure 2.27.

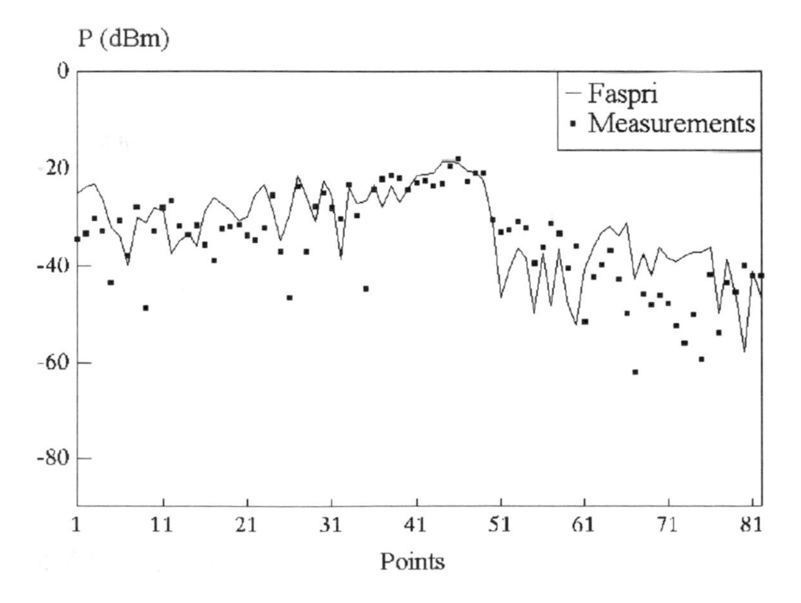

Figure 2.29 Comparison between measurements and simulation for LOS path 1b of Figure 2.27.

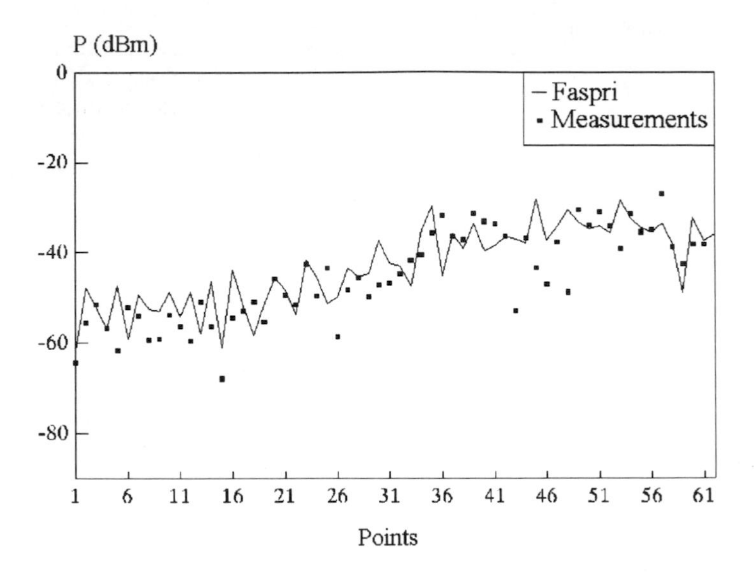

Figure 2.30 Comparison between measurements and simulation corresponding to path 2 of Figure 2.27.

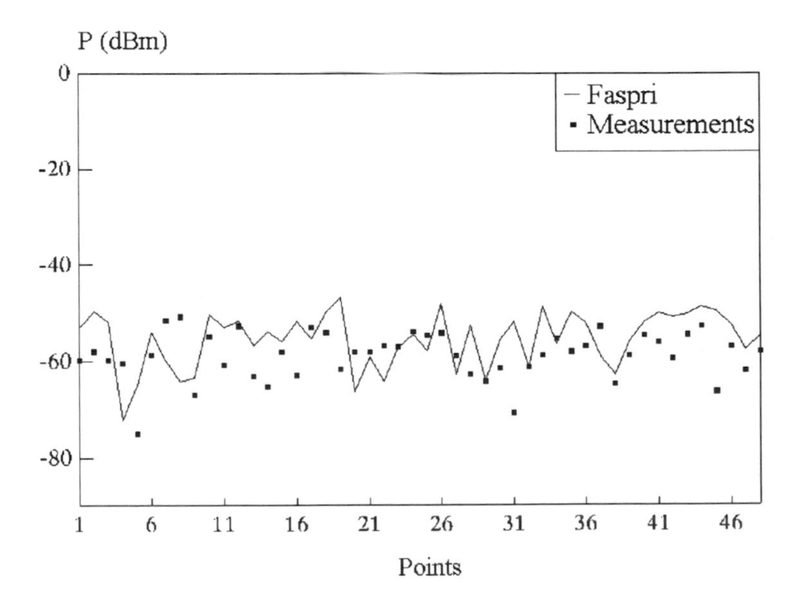

Figure 2.31 Comparison between measurements and simulation for path 3 of Figure 2.27.

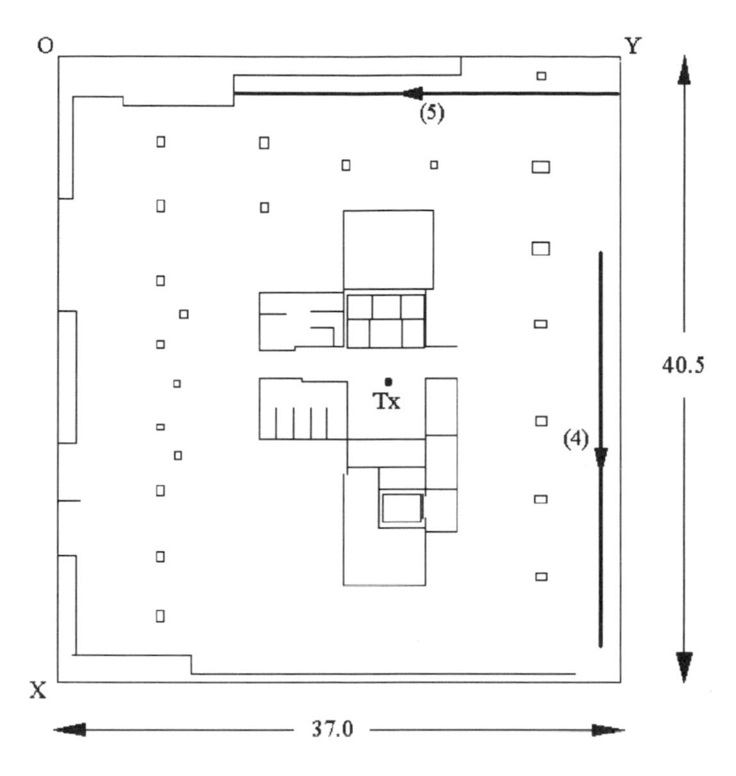

Figure 2.32 Floor plan of the indoor scenario. The bold lines represent the two paths analyzed with the Tx antenna located at (21.0, 21.5, 2.85) and oriented along the –Z axis.

such that they are directly illuminated. It also assumes that the buildings in proximity of the receiver are of uniform height. The field at the receiver is calculated as the sum of two GTD/UTD contributions: a ray diffracted at the last building edge before the receiver (ray 1 in Figure 2.35) and a ray reflected at the next building wall (ray 2 in Figure 2.35). The resulting equations for this model are given in Chapter 4.

Many of the NLOS semi-deterministic models are based on the city configuration initially analyzed by Walfisch and Bertoni [37]. This model of urban scenes is inspired by the typical urban structure in U.S. cities outside of the high-rise core. The buildings are organized along parallel rows with the street lanes between them (see Figure 2.36). The discontinuities between buildings in the same row (passageways) are not taken into account because they tend to be smaller than the building's width and, except when the passageways are aligned with the transmitter from row to row, propagation takes place over the rooftops. For a given street, field coverage at the street level comes mainly from the wave diffracted at the previous rooftop (local rooftop) and

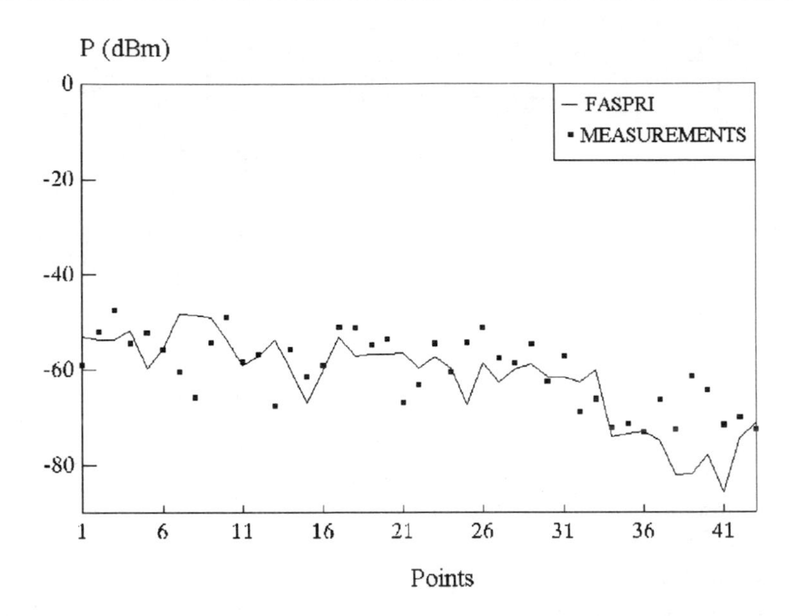

Figure 2.33 Comparison between measurements and simulation for path 4 of Figure 2.32.

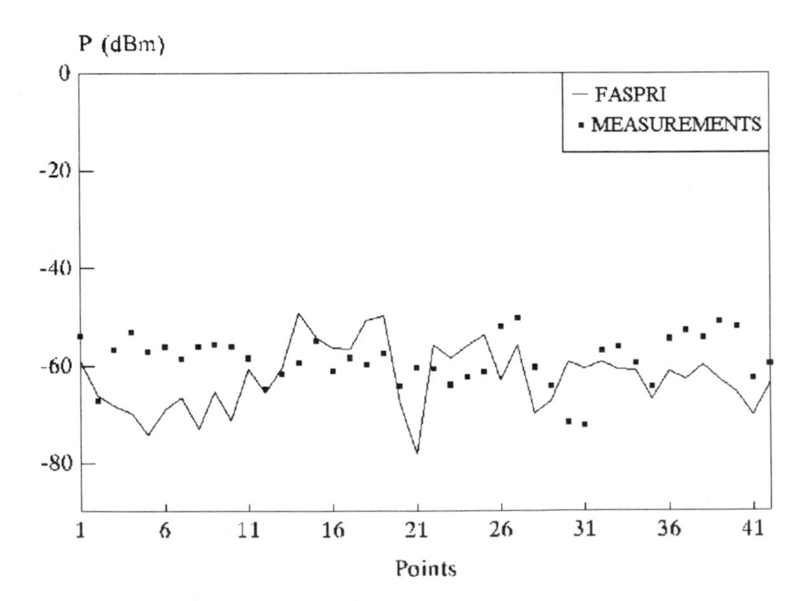

Figure 2.34 Comparison between measurements and simulation for path 5 of Figure 2.32.

Table 2.3
Mean Error Predictions and Standard Deviations for the Six Paths Analyzed

	Mean Error	Standard Deviation
Path 1a	−1.37	7.04
Path 1b	2.57	8.32
Path 2	1.65	7.96
Path 3	3.38	6.79
Path 4	−1.66	7.29
Path 5	−3.59	8.33

from the wave diffracted at the local rooftop and then reflected off a building in the following row. The incident wave at the local rooftop is the result of the forward diffraction in the previous rooftops from the transmitter antenna. Considering that all the buildings in a given row have similar heights and that, at grazing incidence, the shape of the rooftops has low influence on the diffracted wave, the rows of buildings can be substituted by absorbing half-screens (see Figure 2.37).

Therefore, to predict the field at the receiver, one must first calculate the forward diffracted field (multiple knife-edge diffraction) incident to the local half-screen. Then, the received field is calculated due to two contributions: the field diffracted at the local edge and then reflected off the next building. To account for this last reflection, a screen with the electrical properties of the corresponding building should be considered. These two mechanisms, single diffraction and diffraction-reflection, can be treated with the conventional GTD/UTD [38]. However, these resulting signals will be of similar amplitude so, the average field strength can be obtained as twice the single diffracted field.

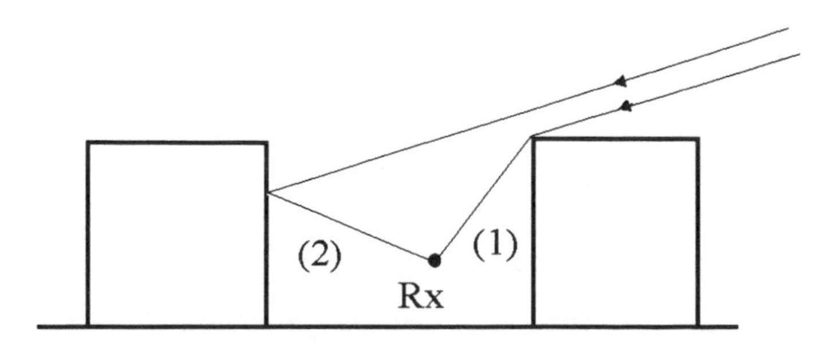

Figure 2.35 The diffracted and reflected rays in the model of Ikegami, et al.

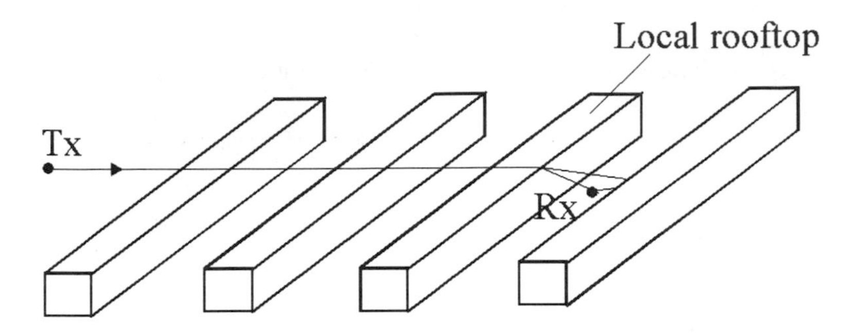

Figure 2.36 Urban model proposed in [37].

To predict the multiple forward diffracted field, PO and GTD techniques have been proposed. The problem with the GTD application is that one screen edge is in the transition zones from the previous screen edges so one cannot apply the UTD diffraction coefficients of Section 2.4. Heuristic modifications to the UTD coefficients have been proposed in [39–42], which overcome the problem of the multiple knife-edge diffractions in the transition zones.

The application of the GTD/UTD in semi-deterministic models is not restricted to NLOS urban propagation. The so-called "two-ray" model has been widely used in rural environments and it has been proposed for microcells with low base station antennas where there is an LOS between transmitter and receiver [25, 43]. In this model the field is calculated as the sum of two contributions: the direct ray and the ground reflected ray.

A more sophisticated model for LOS propagation with a low base station antenna is the "multislit waveguide" model [44]. The streets are modeled as 3D waveguides with randomly distributed slits and screens. The total field along the street is obtained as the sum of the following GTD contributions: direct ray, ground reflected ray, wall reflected rays, and rays diffracted in the building edges.

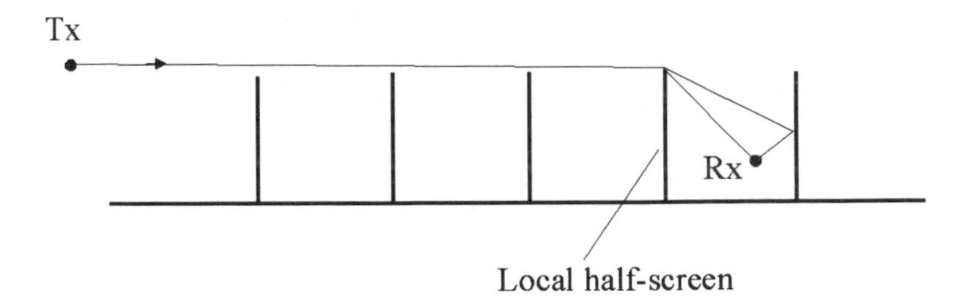

Figure 2.37 Multiple absorbing half-screens urban model.

2.9 Physical Optics

In the context of the radio propagation prediction, the PO approach [5, 6, 45, 46] (namely, the Kirchhoff method, Huygens' method, and tangent plane approximation) has been used in two different problems.

First, it has been applied in urban and rural models to predict multiple diffraction. As stated in previous sections, in some circumstances (usually in semi-deterministic models), the conventional GTD/UTD approach cannot adequately predict the multiple diffraction phenomenon. Hence, many authors have appealed to the PO approach [37, 47–50]. Some other authors have proposed heuristic modifications to the conventional GTD/UTD expressions to overcome its limitations in the multiple diffraction problem (see Section 2.8).

Second, the PO has been widely used in deterministic models to predict the field scattered by the surfaces of the environmental obstacles. The GTD/UTD approach assumes that, in the reflection phenomenon, all the energy is reflected in the specular direction. In the strictest terms, this is true when the surfaces have infinite size and are smooth (not rough). If the surfaces have finite size or are rough, part of the incident wave energy is scattered in directions different from the specular one. Therefore, to overcome this deficiency of the GTD/UTD, the scattering field is included, as an additional mechanism, in various urban deterministic models [23, 24, 27, 51] and rural deterministic models [27, 33, 34]. The contribution of the scattering waves becomes especially important when the receiver is located close to the scattering surface or when there is no first-order GTD/UTD contribution at the receiver.

2.10 Application of Physical Optics in Semi-Deterministic Propagation Models: Multiple Knife-Edge Diffraction

Before its usage in urban models, the multiple half-screen configuration (see Section 2.8) had been widely used in radio propagation prediction in hilly rural areas [52, 53]. In such cases, the hills are modeled by half-screens with the edge (knife edge) located at the apex of the hill ridge. This interesting subject is outside the scope of this book, so hereafter we focus on urban environments.

As mentioned in Section 2.8, the PO approach has been widely used in semi-deterministic models based on the multiple half-screen layout. To estimate the forward diffracted field in such a scenario, various PO-based methods have been proposed. Vogler [54] proposes an expression of the multiple diffraction, which involves a multiple integral whose dimension is equal to the number of edges. This expression results from a repeated application of the Kirchhoff integral to the plane of each screen. This approach is very general in the sense that there are no restrictions on the height of each screen or on the separation between consecutive screens. But the time required to evaluate it is prohibitive even for a low number of diffractions. To overcome this limitation, Vogler expands the integral and approximates it in a suitable form for computer implementation.

Saunders and Bonar [49] have evaluated the Vogler integral using a Monte Carlo numerical method. Apart from that strategy, these authors have obtained a simple propagation model (in a closed-form expression) by approximating the Vogler integral in the case of uniform height and uniform separation between half-screens [48]. They also have developed a heuristic procedure, which combines the above strategies, in order to analyze half-screen configurations with arbitrary height and separations [49].

Walfisch and Bertoni [37] proposed a model based on the direct evaluation of the Kirchhoff–Huygens integral. The field at each screen is obtained from the field diffracted at the previous screen by using a numerical approximation of the integral. The model assumes uniform height and spacing of the screens. Moreover, the transmitter antenna must be above rooftop level. The resulting expressions for this model can be found in Chapter 4.

Xia and Bertoni [47] expressed the incident field at each edge in terms of a multidimensional Fresnel integral similar to the Vogler integral. Then, the integral is expanded into a series of functions (Boersma functions) obtaining a closed-form expression of the field at each edge. The model assumes the screens have uniform height and spacing but there are no limitations on the transmitter antenna height. The resulting expressions of the complete model [38] (including the last diffraction toward the receiver) are given in Chapter 4.

2.11 Application of Physical Optics in Deterministic Propagation Models: Scattering Field

Unlike the GTD approach, the PO technique is not a ray-based technique so the scattering problems must be posed in terms of induced currents instead of in terms of rays. When an incident wave reaches a scattering surface, a surface current distribution (magnetic and electric) is induced. These currents are responsible for the scattered field. Rigorous determination of the induced currents is a complicated problem. (In fact, the scattering phenomenon is one of the most important topics in the electromagnetic context). The PO provides approximate, but simple, expressions of the induced currents. The PO scattering equations can be derived using the Equivalent Principle [5] and applying the so-called Tangent Plane approximation [5, 46]. The derivation of the PO expressions is outside the scope of this book so instead we focus on the resulting PO equations and their application to the scattering problem in the radio propagation context.

The PO expresses the induced currents as functions of the incident field at the scattering surface. PO assumes currents to be zero at points not illuminated by the incident wave and the following currents at illuminated points:

$$\vec{J}(\vec{r}') = \hat{n} \times \begin{bmatrix} 1 + \Gamma_s & 0 \\ 0 & 1 + \Gamma_h \end{bmatrix} \begin{bmatrix} H_s^i(\vec{r}') \\ H_h^i(\vec{r}') \end{bmatrix} \tag{2.53}$$

$$\vec{M}(\vec{r}') = -\hat{n} \times \begin{bmatrix} 1 + \Gamma_s & 0 \\ 0 & 1 + \Gamma_h \end{bmatrix} \begin{bmatrix} E_s^i(\vec{r}') \\ E_h^i(\vec{r}') \end{bmatrix} \tag{2.54}$$

where $\hat{n}$ is the normal unit vector at the surface point $\vec{r}'$, $\Gamma_{s,h}$ are the Fresnel reflection coefficients (see (2.15) and (2.16)), and $H_{s,h}^i(\vec{r}')$ and $E_{s,h}^i(\vec{r}')$ are the soft and hard components of the incident magnetic and electric field vectors at the surface point, respectively.

Because PO predicts currents of zero at points not illuminated by the incident wave, only the illuminated part of the surface will contribute to the scattering field. Moreover, only the part of the surface visible from the receiver will contribute to the scattering field.

To evaluate the field contribution of the scattering surface, one must calculate the field radiated by the induced currents of (2.53) and (2.54). This is not an easy task so, in practice, to simplify the problem, the surface is segmented into small, but electrically large, plane facets (typically rectangular squares) [23, 24, 27, 33]. The individual contributions of the resulting facets are added to obtain the total scattering field. When the facets are viewed from the incident wave source and from the observation point under small solid angles, two assumptions can be made: The incident wave at the surface is a plane and the scattered wave at the receiver is also a plane wave. Under these assumptions, the problem is similar to a bistatic radar cross-section (RCS) problem. Then, assuming a smooth facet, the scattering electric field at the receiver ($\vec{E}_s$) can be obtained as:

$$\vec{E}_s = \vec{E}_J + \vec{E}_M \tag{2.55}$$

where

$$\vec{E}_J = \frac{jf\mu}{2} \frac{\exp(-j\beta r)}{r} \hat{r} \times \left(\hat{r} \times \int_S \vec{J}(\vec{r}') \exp(j\beta \hat{r} \cdot \vec{r}')\, ds \right) \tag{2.56}$$

$$\vec{E}_M = \frac{jf\mu\epsilon}{2} \frac{\exp(-j\beta r)}{r} \hat{r} \times \int_S \vec{M}(\vec{r}') \exp(j\beta \hat{r} \cdot \vec{r}')\, ds \tag{2.57}$$

where r is the distance between the center of the facet and the observation point, $\hat{r}$ is a unit vector from the center of the facet to the observation point, f is the frequency, β is the wave number, S is the surface of the facet, and ϵ and μ are the permittivity and the permeability of the free space, respectively. Assuming that the incident wave is plane at the facet and considering (2.53) and (2.54), the preceding expressions can be written as

$$\vec{E}_J = \frac{jf\mu}{2} \frac{\exp(-j\beta r)}{r} \hat{r} \times (\hat{r} \times [\hat{n} \times \vec{H}^1])\, I_{\mathrm{PO}} \tag{2.58}$$

$$\vec{E}_M = \frac{-jf\mu\epsilon}{2} \frac{\exp(-j\beta r)}{r} \hat{r} \times (\hat{n} \times \vec{E}^1)\, I_{\mathrm{PO}} \tag{2.59}$$

where $\hat{n}$ is the normal vector to the facet, I_{PO} is the so-called PO integral, and $\vec{H}_1$ and $\vec{E}_1$ are constant vectors given by

$$\begin{bmatrix} H_s^1 \\ H_h^1 \end{bmatrix} = \begin{bmatrix} 1 + \Gamma_s & 0 \\ 0 & 1 + \Gamma_h \end{bmatrix} \begin{bmatrix} H_s^0 \\ H_h^0 \end{bmatrix} \tag{2.60}$$

$$\begin{bmatrix} E_s^1 \\ E_h^1 \end{bmatrix} = \begin{bmatrix} 1 + \Gamma_s & 0 \\ 0 & 1 + \Gamma_h \end{bmatrix} \begin{bmatrix} E_s^0 \\ E_h^0 \end{bmatrix} \tag{2.61}$$

with $\vec{H}^0$ and $\vec{E}^0$ being the amplitudes of the electric and magnetic fields associated with the incident plane wave at the facet.

The PO integral is given by

$$I_{\mathrm{PO}} = \int_S \exp[j\beta(\hat{r} - \hat{r}_i) \cdot \vec{r}']\, ds \tag{2.62}$$

where $\hat{r}_i$ is the incident direction of the plane wave toward the facet. The PO integral can be written in a closed-form expression [55]. Considering a local coordinate system with the Z axis parallel to $\hat{n}$ and the origin in a vertex of the facet, I_{PO} can be expressed as follows:

$$I_{PO} = \sum_{n=1}^{N} \frac{\hat{y} \cdot (\vec{a}_{n+1} - \vec{a}_n)}{K_x} \operatorname{sinc}\left[\frac{\vec{K} \cdot (\vec{a}_{n+1} - \vec{a}_n)}{\pi}\right] \exp\left[\frac{j\vec{K} \cdot (\vec{a}_{n+1} + \vec{a}_n)}{2}\right]; \quad \text{if } K_x \neq 0$$

$$I_{PO} = \sum_{n=1}^{N} \frac{-\hat{x} \cdot (\vec{a}_{n+1} - \vec{a}_n)}{K_y} \operatorname{sinc}\left[\frac{\vec{K} \cdot (\vec{a}_{n+1} - \vec{a}_n)}{\pi}\right] \exp\left[\frac{j\vec{K} \cdot (\vec{a}_{n+1} + \vec{a}_n)}{2}\right]; \quad \text{if } K_y \neq 0$$

$$= AREA; \quad \text{if } K_x = K_y = 0 \tag{2.63}$$

where $\vec{a}_i$ are the vertices of the facet (referring to the local coordinate system), N is the number of vertices, $AREA$ is the area of the facet (see Appendix 3A) and

$$\vec{K} = \frac{\beta(\hat{r} - \hat{r}_i)}{2} \tag{2.64}$$

This expression assumes that the scattering surface is smooth. The effects of its roughness can be easily included by multiplying the Fresnel coefficients by the scattering loss factor as was pointed out in Section 2.3.

References

[1] Pahlavan, K., and A. H. Levesque, *Wireless Information Networks,* New York: John Wiley, 1995.

[2] Conde, O. M., and M. F. Cátedra, "Iterative Technique for Scattering and Propagation Over Arbitrary Environment," 13th Annual Review of Progress in Applied Computational Electromagnetics (ACES), Monterey, CA, March 1997, pp. 1310–1317.

[3] Hviid, J. T., J. Bach Andersen, J. Toftgard, and J. Bojer, "Terrain-Based Propagation Model for Rural Area–An Integral Equation Approach," *IEEE Trans. on Antennas and Propagation,* Vol. AP-43, January 1995, pp. 41–46.

[4] McNamara, D. A., C. W. I. Pistorious, and J. A. G. Maherbe, *Introduction to the Uniform Geometric Theory of Diffraction,* Norwood, MA: Artech House, 1990.

[5] Balanis, C. A., *Advanced Engineering Electromagnetics,* New York: John Wiley, 1989.

[6] James, G. L., *Geometrical Theory of Diffraction for Electromagnetic Waves,* 3rd ed., London: Peter Peregrinus Ltd., 1986.

[7] Kouyoumjian, R. G., "The Geometrical Theory of Diffraction and Its Application," Chap. 6 in *Numerical and Asymptotic Techniques in Electromagnetics,* R. Mittra, Ed., New York: Springer, 1975.

[8] Kouyoumjian, R. G., and P. H. Pathak, "A Uniform Geometrical Theory of Diffraction for an Edge in a Perfectly Conducting Surface," *Proc. IEEE,* Vol. 62, No. 11, Nov. 1974, pp. 1448–1461.

[9] Sou, C. K., O. Landron, and M. J. Feuerstein, "Characterization of Electromagnetic Properties of Building Materials for Use in Site-Specific Propagation Prediction," MPRG Technical Report No. 92-12, Virginia Polytechnic and State University, Blacksburg, 1992.

[10] Landron, O., M. J. Fuerstein, and T. S. Rappaport, "A Comparison of Theoretical and Empirical Reflection Coefficients for Typical Exterior Wall Surfaces in a Mobile Radio Environment," *IEEE Trans. on Antennas and Propagation,* Vol. 44, No. 3, March 1996, pp. 341–351.

[11] Boithias, L., *Radio Wave Propagation,* New York: McGraw-Hill, 1987.

[12] Rappaport, T. S., *Wireless Communications. Principles and Practice,* Upper Saddle River, NJ: Prentice Hall, 1996.

[13] Stratton, J. A., *Electromagnetic Theory,* New York: McGraw-Hill, 1941.

[14] Chew, W. C., *Waves and Fields in Inhomogeneous Media,* New York: IEEE Press, 1995.

[15] Schneider, M., and R. J. Luebbers, "A General, Uniform Double Wedge Diffraction Coefficient," *IEEE Trans. on Antennas and Propagation,* Vol. 39, No. 1, January 1991, pp. 8–14.

[16] Seidel, S. Y., and T. S. Rappaport, "Site-Specific Propagation Prediction for Wireless in Building Personal Communication System Design," *IEEE Trans. on Vehicular Technology,* Vol. 43, No. 4, Nov. 1994, pp. 879–891.

[17] Honcharenko, W., H. L. Bertoni, J. L. Dailing, J. Quian, and H. D. Yee, "Mechanism Governing UHF Propagation on Single Floors in Modern Office Buildings," *IEEE Trans. on Vehicular Technology,* Vol. 41, No. 4, November 1992, pp. 496–504.

[18] Holt, T., K. Pahlavan, and J. F. Lee, "A Graphical Indoor Radio Channel Simulator Using 2D Ray Tracing," Proc. PIMRC 92, Boston, MA, USA, Oct. 1992, pp. 411–416.

[19] Yang, G., K. Pahlavan, and J. F. Lee, "A 3D Propagation Model With Polarization Characteristics in Indoor Radio Channels," Proc. IEEE GLOBECOM 93, Houston, TX, Nov. 1993.

[20] Tarng, J. H., W. R. Chang, and B. J. Hsu, "Three-Dimensional Modeling of 900-Mhz and 2.44-Ghz Radio Propagation in Corridors," *IEEE Trans. on Vehicular Technology,* Vol. 46, No. 2, May 1997, pp. 519–527.

[21] Lawton, M. C., and J. P. McGeehan, "The Application of a Deterministic Ray Launching Algorithm for the Prediction of Radio Channel Characteristics in Small-Cell Environments," *IEEE Trans. on Vehicular Technology,* Vol. 43, No. 4, November 1994, pp. 955–969.

[22] McKown, J. W., and R. L. Hamilton, "Ray Tracing as a Design Tool for Radio Networks," *IEEE Network Mag.,* November 1991, pp. 27–30.

[23] Rappaport, T. S., S. Y. Seidel, and K. R. Schaubach, "Site-Specific Propagation for PCS System Design," in *Wireless Personal Communications,* M. J. Feuerstein, T. S. Rappaport, Eds., Boston: Kluwer Academic Publishers, 1993, pp. 281–315.

[24] Schaubach, K. R., and N. J. Davis, "Microcellular Radio-Channel Propagation Prediction," *IEEE Antennas and Propagation Mag.,* Vol. 36, No. 4, Aug. 1994, pp. 25–34.

[25] Rustako, A. J., N. Amitay, G. J. Owens, and R. S. Roman, "Radio Propagation at Microwave Frequencies for Line-of-Sight Microcellular Mobile and Personal Communications," *IEEE Trans. on Vehicular Technology,* Vol. 40, 1991, pp. 203–210.

[26] Tan, S. Y., and H. S. Tan, "A Microcellular Communications Propagation Model Based on the Uniform Theory of Diffraction and Multiple Image Theory," *IEEE Trans. on Antennas and Propagation,* Vol. 44, No. 10, Oct. 1996, pp. 1317–1326.

[27] Kürner, T., D. J. Cichon, and W. Wiesbeck, "Concepts and Results for 3D Digital Terrain-Based Wave Propagation Models: An Overview," *IEEE J. Selected Areas in Communications,* Vol. 11, No. 7, Sep. 1993, pp. 1002–1012.

[28] Cátedra, M. F., J. Pérez, A. González, O. Gutiérrez, and F. Saez de Adana, "Fast Computer Tool for the Analysis of Propagation in Urban Cells," Proc. of Wireless Communications Conference, Boulder, CO, Aug. 11–13, 1997, pp. 240–245.

[29] Cátedra, M. F., J. Pérez, F. Saez de Adana, and O. Gutiérrez, "Efficient Ray-Tracing Techniques for 3D Analysis of Propagation in Mobile Communications. Application to Picocell and Microcell Scenarios," *IEEE Antennas and Propagation Mag.,* Vol. 40, No. 2, April 1998, pp. 15–28.

[30] Kanatas, A. G., I. D. Kountouris, G. B. Kostaras, and P. Constantinou, "A UTD Propagation Model in Urban Microcellular Environments," *IEEE Trans. on Vehicular Technology,* Vol. 46, No. 1, February 1997.

[31] Luebbers, R. J., "Propagation Prediction for Hilly Terrain Using GTD Wedge Diffraction," *IEEE Trans. on Antennas and Propagation,* Sep. 1984, pp. 951–955.

[32] Luebbers, R. J., W. A. Foose, and G. Reyner, "Comparison of GTD Propagation Model for Wide-Band Path Loss Simulation With Measurements," *IEEE Trans. on Antennas and Propagation,* Vol. 37, No. 4, Apr. 1989, pp. 499–505.

[33] Lebherz, M., W. Wiesbeck, and W. Krank, "A Versatile Wave Propagation Model for the VHF/UHF Range Considering Three-Dimensional Terrain," *IEEE Trans. on Antennas and Propagation,* Vol. 40, No. 10, Oct. 1992, pp. 1121–1131.

[34] Kürner, T., D. J. Cichon, and W. Wiesbeck, "Evaluation and Verification of the VHF/UHF Propagation Channel Based on a 3-D-Wave Propagation Model," *IEEE Trans. on Antennas and Propagation,* Vol. 44, No. 3, March 1996.

[35] Cátedra, M. F., J. Pérez, F. S. de Adana, O. Gutiérrez, J. Cantalapiedra, and I. González, "Fast ray-tracing method for calculating the propagation in indoor environments," Paper submitted to IEEE AP-S International Symposium, Atlanta, Georgia, USA, June 21–26, 1998.

[36] Ikegami, F., S. Yoshida, T. Takeuchi, and M. Umehira, "Propagation Factors Controlling Mean Field Strength on Urban Streets," *IEEE Trans. on Antennas and Propagation,* Vol. 32, No. 12, Dec. 1984, pp. 822–829.

[37] Walfisch, J., and H. L. Bertoni, "A Theoretical Model of UHF Propagation in Urban Environments," *IEEE Trans. on Antennas and Propagation,* Vol. 36, No. 12, Dec. 1988. pp. 1788–1796.

[38] Bertoni, H. L., W. Honcharenko, L. R. Maciel, and H. H. Xia, "UHF Propagation Prediction for Wireless Personal Communications," *Proc. IEEE,* Vol. 82, No. 9, Sep. 1994, pp. 1333–1358.

[39] Holm, P. D., "UTD-Diffraction Coefficients for Higher Order Wedge Diffracted Fields," *IEEE Trans. on Antennas and Propagation,* Vol. 44, No. 6, June 1996, pp. 879–888.

[40] Andersen, J. B., "UTD Multiple-Edge Transition Zone Diffraction," *IEEE Trans. on Antennas and Propagation,* Vol. 45, No. 7, July 1997.

[41] Zhang, W., "A Wide-Band Propagation Model Based on UTD for Cellular Mobile Radio Communications," *IEEE Trans. on Antennas and Propagation,* Vol. 45, No. 11, Nov. 1997, pp. 1669–1678.

[42] Zhang, W., "A UTD-Based Propagation Model for Cellular Mobile Radio Systems," Proc. 25th European Microwave Conference, Bologna, Italy, Sep. 1995, pp. 276–280.

[43] Xia, H. H., H. L. Bertoni, L. R. Maciel, A. Lindsay-Steward, and R. Rowe, "Radio Propagation Characteristics for Line-of-Sight Microcellular and Personal Communications," *IEEE Trans. on Antennas and Propagation,* Vol. 41, No. 10, 1993, pp. 1439–1447.

[44] Blaunstein, N., and M. Levin, "VHF/UHF Wave Attenuation in a City With Regularly Spaced Buildings," *Radio Science,* Vol. 31, No. 2, Mar./Apr. 1996, pp. 313–323.

[45] Felsen, L. B., and N. Marcuvitz, *Radiation and Scattering of Waves,* Englewood Cliffs, NJ: Prentice Hall, 1973.

[46] Ruck, G. T., D. E. Barrick, W. D. Stuart, and D. K. Krichbaum, *Radar Cross Section Handbook,* New York: Plenum, 1970.

[47] Xia, H. H., and H. L. Bertoni, "Diffraction of Cylindrical and Plane Waves by an Array of Absorbing Half-Screens," *IEEE Trans. on Antennas and Propagation,* Vol. 40, No. 2, Feb. 1992, pp. 170–177.

[48] Saunders, S. R., and F. R. Bonar, "Explicit Multiple Building Diffraction Attenuation Function for Mobile Radio Wave Propagation," *Electronic Letters,* Vol. 27, 1991, 1276–1277.

[49] Saunders, S. R., and F. R. Bonar, "Prediction of Mobile Radio Wave Propagation Over Buildings of Irregular Heights and Spacing," *IEEE Trans. on Antennas and Propagation,* Vol. 42, No. 2, Feb. 1994, pp. 137–144.

[50] Russell, T. A., C. W. Bostian, and T. S. Rappaport, "A Deterministic Approach to Predicting Microwave Diffraction by Buildings for Microcellular Systems," *IEEE Trans. on Antennas and Propagation,* Vol. 41, No. 12, Dec. 1993, pp. 1640–1649.

[51] Al-Nuaimi, M. O., and M. S. Ding, "Prediction Models and Measurements of Microwave Signals Scattered From Buildings," *IEEE Trans. on Antennas and Propagation,* Vol. 42, No. 8, Aug. 1994, pp. 1126–1137.

[52] Bullington, K., "Radio Propagation at Frequencies Above 30 Megacycles," *Proc. IEEE,* Vol. 35, 1947, pp. 1122–1136.

[53] Lee, W. C. Y., *Mobile Communications Engineering,* New York: McGraw-Hill, 1985.

[54] Vogler, L. E., "An Attenuation Function for Multiple Knife-Edge Diffraction," *Radio Science,* Vol. 17, No. 6, Nov./Dec. 1982, pp. 1541–1546.

[55] Gordon, W. B., "Far-Field Approximation to the Kirchhoff-Helmholtz Representations of Scattered Field," *IEEE Transaction on Antennas and Propagation,* Vol. 23, July 1975, pp. 864–876.

Appendix 2A: Vector Transformations

2A.1 Vector Transformations Between Rectangular and Spherical Coordinate Systems

The description of a vector in a rectangular coordinate system is $\vec{v} = x\hat{a}_x + y\hat{a}_y + z\hat{a}_z$ where the values (x, y, z) are known as the cartesian or rectangular vector coordinates (see Figure 2A.1).

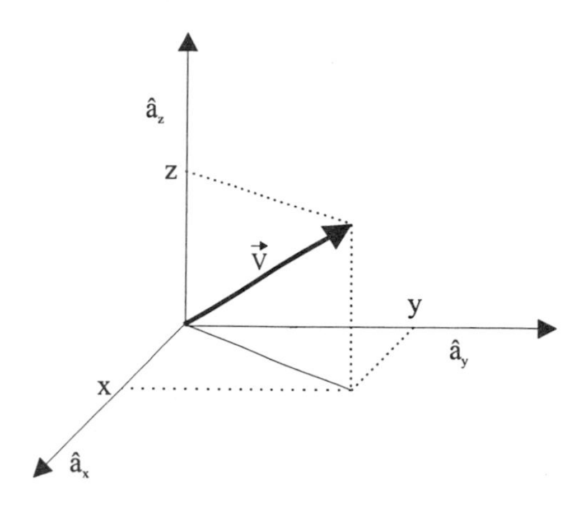

Figure 2A.1 Rectangular coordinate system.

The description of a vector in a spherical coordinate system is $\vec{v} = r\hat{a}_r + \theta\hat{a}_\theta + \phi\hat{a}_\theta$ where the values (r, θ, ϕ) are known as the spherical vector coordinates (see Figure 2A.2).

The relationships between the unit vectors of the rectangular and spherical coordinate systems are:

$$\hat{a}_x = \sin\theta\cos\phi\ \hat{a}_r + \cos\theta\cos\phi\ \hat{a}_\theta - \sin\phi\ \hat{a}_\theta \tag{2A.1}$$

$$\hat{a}_y = \sin\theta\sin\phi\ \hat{a}_r + \cos\theta\sin\phi\ \hat{a}_\theta + \cos\phi\ \hat{a}_\phi \tag{2A.2}$$

$$\hat{a}_z = \cos\theta\ \hat{a}_r - \sin\theta\ \hat{a}_\theta \tag{2A.3}$$

$$\hat{a}_r = \sin\theta\cos\phi\ \hat{a}_x + \sin\theta\sin\phi\ \hat{a}_y + \cos\theta\ \hat{a}_z \tag{2A.4}$$

$$\hat{a}_\theta = \cos\theta\cos\phi\ \hat{a}_x + \cos\theta\sin\phi\ \hat{a}_y - \sin\theta\ \hat{a}_z \tag{2A.5}$$

$$\hat{a}_\phi = -\sin\phi\ \hat{a}_x + \cos\phi\ \hat{a}_y \tag{2A.6}$$

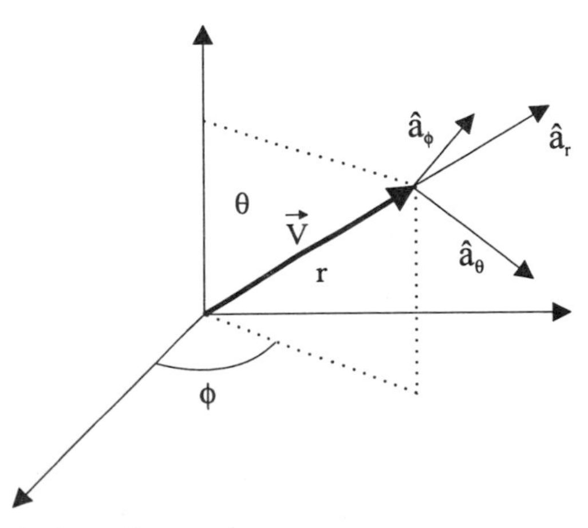

Figure 2A.2 Spherical coordinate system.

Therefore, the vector transformation from the rectangular to the spherical coordinate systems is given by

$$\begin{pmatrix} r \\ \theta \\ \phi \end{pmatrix} = \begin{pmatrix} \sin\theta\cos\phi & \sin\theta\sin\phi & \cos\theta \\ \cos\theta\cos\phi & \cos\theta\sin\phi & -\sin\theta \\ -\sin\phi & \cos\phi & 0 \end{pmatrix} \begin{pmatrix} x \\ y \\ z \end{pmatrix} \quad (2A.7)$$

and the vector transformation from the spherical to the rectangular coordinate systems is given by:

$$\begin{pmatrix} x \\ y \\ z \end{pmatrix} = \begin{pmatrix} \sin\theta\cos\phi & \cos\theta\cos\phi & -\sin\phi \\ \sin\theta\sin\phi & \cos\theta\sin\phi & \cos\phi \\ \cos\theta & -\sin\theta & 0 \end{pmatrix} \begin{pmatrix} r \\ \theta \\ \phi \end{pmatrix} \quad (2A.8)$$

2A.1.1 Vector Transformations Between Rectangular Coordinate Systems

Let's consider two rectangular coordinate systems with unit vectors $\hat{a}_x$, $\hat{a}_y$, $\hat{a}_z$ and $\hat{a}_x'$, $\hat{a}_y'$, $\hat{a}_z'$ as Figure 2A.3 shows. Let's consider a vector $\hat{v}$ with coordinates (x, y, z) and (x', y', z') in these coordinate systems:

$$\vec{v} = x\hat{a}_x + y\hat{a}_y + z\hat{a}_z, \qquad \vec{v} = x'\hat{a}_x' + y'\hat{a}_y' + z'\hat{a}_z' \quad (2A.9)$$

The relationships between the unit vectors of the coordinate systems are

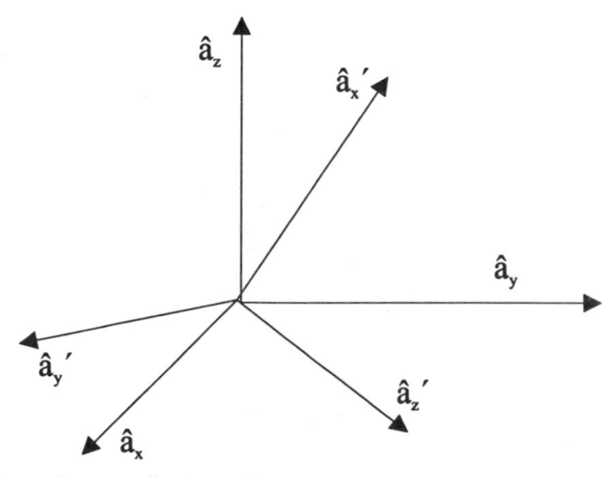

Figure 2A.3 Rectangular coordinate systems.

$$\hat{a}_x' = (\hat{a}_x' \cdot \hat{a}_x)\hat{a}_x + (\hat{a}_x' \cdot \hat{a}_y)\hat{a}_y + (\hat{a}_x' \cdot \hat{a}_z)\hat{a}_z \tag{2A.10}$$

$$\hat{a}_y' = (\hat{a}_y' \cdot \hat{a}_x)\hat{a}_x + (\hat{a}_y' \cdot \hat{a}_y)\hat{a}_y + (\hat{a}_y' \cdot \hat{a}_z)\hat{a}_z \tag{2A.11}$$

$$\hat{a}_z' = (\hat{a}_z' \cdot \hat{a}_x)\hat{a}_x + (\hat{a}_z' \cdot \hat{a}_y)\hat{a}_y + (\hat{a}_z' \cdot \hat{a}_z)\hat{a}_z \tag{2A.12}$$

$$\hat{a}_x = (\hat{a}_x' \cdot \hat{a}_x)\hat{a}_x' + (\hat{a}_y' \cdot \hat{a}_x)\hat{a}_y' + (\hat{a}_z' \cdot \hat{a}_x)\hat{a}_z' \tag{2A.13}$$

$$\hat{a}_y = (\hat{a}_x' \cdot \hat{a}_y)\hat{a}_x' + (\hat{a}_y' \cdot \hat{a}_y)\hat{a}_y' + (\hat{a}_z' \cdot \hat{a}_y)\hat{a}_z' \tag{2A.14}$$

$$\hat{a}_z = (\hat{a}_x' \cdot \hat{a}_z)\hat{a}_x' + (\hat{a}_y' \cdot \hat{a}_z)\hat{a}_y' + (\hat{a}_z' \cdot \hat{a}_z)\hat{a}_z' \tag{2A.15}$$

The dot products $(\hat{a}_i' \cdot \hat{a}_j)$ are known as the director cosines. They are the cosines of the angles formed by the unit vectors of both coordinate systems. Considering equations 2A9–15, the vector transformation from the coordinate system $(\hat{a}_x, \hat{a}_y, \hat{a}_z)$ to the coordinate system $(\hat{a}_x', \hat{a}_y', \hat{a}_z')$ is given by

$$\begin{pmatrix} x' \\ y' \\ z' \end{pmatrix} = \begin{pmatrix} \hat{a}_x' \cdot \hat{a}_x & \hat{a}_x' \cdot \hat{a}_y & \hat{a}_x' \cdot \hat{a}_z \\ \hat{a}_y' \cdot \hat{a}_x & \hat{a}_y' \cdot \hat{a}_y & \hat{a}_y' \cdot \hat{a}_z \\ \hat{a}_z' \cdot \hat{a}_x & \hat{a}_z' \cdot \hat{a}_y & \hat{a}_z' \cdot \hat{a}_z \end{pmatrix} \begin{pmatrix} x \\ y \\ z \end{pmatrix} \tag{2A.16}$$

and the vector transformation from the coordinate system $(\hat{a}_x', \hat{a}_y', \hat{a}_z')$ to the coordinate system $(\hat{a}_x, \hat{a}_y, \hat{a}_z)$ is given by

$$\begin{pmatrix} x \\ y \\ z \end{pmatrix} = \begin{pmatrix} \hat{a}_x' \cdot \hat{a}_x & \hat{a}_y' \cdot \hat{a}_x & \hat{a}_z' \cdot \hat{a}_x \\ \hat{a}_x' \cdot \hat{a}_y & \hat{a}_y' \cdot \hat{a}_y & \hat{a}_z' \cdot \hat{a}_y \\ \hat{a}_x' \cdot \hat{a}_z & \hat{a}_y' \cdot \hat{a}_z & \hat{a}_z' \cdot \hat{a}_z \end{pmatrix} \begin{pmatrix} x' \\ y' \\ z' \end{pmatrix} \tag{2A.17}$$

2A.1.2 Vector Transformations Between Spherical Coordinate Systems

The vector transformation from a spherical coordinate $(\hat{a}_r, \hat{a}_\theta, \hat{a}_\phi)$ to another spherical coordinate system $(\hat{a}_r', \hat{a}_\theta', \hat{a}_\phi')$ can be achieved indirectly using the vector transformations of the above sections. Firstly, the vector is transformed from the spherical coordinate system $(\hat{a}_r, \hat{a}_\theta, \hat{a}_\phi)$ to the corresponding rectangular coordinate system $(\hat{a}_x, \hat{a}_y, \hat{a}_z)$ by using (2A.8). Secondly, the vector is transformed to the rectangular coordinate system $(\hat{a}_x', \hat{a}_y', \hat{a}_z')$ associated with the final spherical coordinate system $(\hat{a}_r', \hat{a}_\theta', \hat{a}_\phi')$ using (2A.16). Finally the vector is transformed to the final spherical coordinate system applying (2A.7). This procedure is sumarized in Figure 2A.4.

$$r, \theta, \phi \xrightarrow[\text{eq. a.2.8}]{} x, y, z \xrightarrow[\text{eq. a.2.16}]{} x', y', z' \xrightarrow[\text{eq. a.2.7}]{} r', \theta', \phi'$$

Figure 2A.4 Vector transformation between spherical coordinate systems.

Appendix 2B: Fresnel Transition Function

The Fresnel transition function is defined as

$$F[x] = 2j\sqrt{x}e^{jx}\int_{\sqrt{x}}^{\infty}\exp(-j\tau^2)\,d\tau \tag{2B.1}$$

For positive values of x, the above expression can be evaluated approximately as follows

—For $x > 5.5$:

$$F[x] = 1 + \frac{j}{2x} - \frac{3}{4x^2} - \frac{j15}{8x^3} + \frac{75}{16x^4} \tag{2B.2}$$

—For values $0.3 \leq x \leq 5.5$, one can use an interpolation scheme based on the values of Table 2B.1.

—For $0 \leq x \leq 0.3$:

$$F[x] = \left(\sqrt{\pi x} - 2xe^{j\pi/4} - \frac{2x^2 e^{-j\pi/4}}{3}\right)e^{j(x+\pi/4)} \tag{2B.3}$$

—For $x < 0$:

$$F[x] = F^*[|x|] \tag{2B.4}$$

where the asterisk indicates a complex conjugate.

Table 2B.1
Data for Interpolating Values of the Fresnel Transition Function

x	*F*[*x*]
0.3	0.5729 + *j* 0.2677
0.5	0.6768 + *j* 2682
0.7	0.7439 + *j* 0.2549
1.0	0.8095 + *j* 0.2322
1.5	0.8730 + *j* 0.1982
2.3	0.9240 + *j* 0.1577
4.0	0.9658 + *j* 0.1073
5.5	0.9797 + *j* 0.0828

3

Ray-Tracing Techniques

As stated in Chapter 2, the GTD field contribution is calculated as the sum of the fields associated with the rays that reach the observation points. In complex environments, such as 3D scenarios, the principal difficulty of the GTD application is to solve ray-tracing, which consumes most of the simulation time. Therefore, the efficiency of a GTD-based tool depends highly on the ray-tracing simulator.

This chapter deals with ray-tracing techniques suitable for GTD-based deterministic propagation models. First, an adequate method to model such environments is presented: the faceted model. Second, the ray-tracing problem in such scenarios is analyzed. The next section addresses ray-tracing propagation models. Efficient ray-tracing acceleration techniques suitable for this problem are presented later. Finally, rigorous algorithms to achieve ray-tracing are shown.

3.1 Geometrical and Morphological Models

The information required by any deterministic propagation tool can be classified into two types: (1) geometric description of the scene and (2) morphological description of the scene, that is, properties of the object materials of the scene.

3.1.1 Geometric Description

From a geometrical point of view, the typical microcell and picocell scenarios are quite complex in that a number of different objects are involved: buildings, lamp posts, telephone booths, trees, furniture, etc. Moreover, some of them are mobile objects such as cars and persons; all of which are involved in the radio propagation phenomenon, although each has a different influence. As a consequence, without certain simplifications, radio propagation in such environments is impossible to simulate even using approximate electromagnetic techniques. On the other hand, available data usually do not contain geometric information for small objects. Evidently, there is not enough information relative to mobile obstacles.

Therefore, the level of detail in the geometric models must be related to the available data and to the electromagnetic approach used. Consequently, only building data and sometimes terrain data are considered. Sometimes, information about indoor objects is available although this is not the usual case.

In many cases, the only available information about buildings is related to their external walls (geometric shape and location), and sometimes includes information relative to the materials in the walls. This could be enough for outdoor propagation prediction but not for indoor prediction, which requires, at the very least, information about the internal building structure (walls, floors, etc). Information relative to windows and doors is also of interest for indoor models.

Terrain data must be taken into account in the geometric models, especially in small urbanized outdoor environments and in hilly areas where flat ground cannot be assumed. In other cases, microcell and picocell ground can be considered flat. Data about the ground material is always important in outdoor environments.

Building data can be obtained from local governments, architect blueprints, city planners, etc. Currently, the main source of terrain data (in digital form) is the archived records of government agencies. In the United States, for example, the U.S. Geological Survey (USGS), the Environmental Protection Agency (EPA), the National Aeronautics and Space Administration (NASA), and the National Oceanographic and Atmospheric Administration (NOAA) have developed terrain databases that are available to the public at a moderate cost. In the United Kingdom, a partially similar position exists with the British Geological Survey, the Soil Survey, and the Institute of Terrestrial Ecology. In Spain, digital terrain data can be obtained from Centro de Gestión Catastral. At this moment, digital terrain databases are increasingly being generated and distributed by private publishing companies as well.

3.1.2 Morphological Models

Reflective properties of ground and surface building materials must be considered in order to enhance the accuracy of the predictions. They can be obtained from measurements [1–3] or they can be calculated [2, 4] from the electrical properties and roughness of the corresponding materials (see Chapter 2). The electrical properties are:

- Relative permittivity, ϵ_r;
- Relative permeability, μ_r; and
- Conductance, σ.

Tables of electrical properties of materials can be found in the literature [1, 5–7]. They vary with frequency. As an example, typical electric parameters at 1.8 GHz for different types of common building exterior surfaces are [1]:

Limestone: $\epsilon_r' = 7.68$, $\epsilon_r'' = 0.21$, $\mu_r' = 0.96$, $\mu_r'' = 0.006$, $\sigma = 0.03$ S/m,
Brick: $\epsilon_r' = 4.26$, $\epsilon_r'' = 0.09$, $\mu_r' = 1.03$, $\mu_r'' = 0.03$, $\sigma = 0.01$ S/m, and
Concrete: $\epsilon_r' = 6.05$, $\epsilon_r'' = 1.64$, $\mu_r' = 0.95$, $\mu_r'' = -0.05$, $\sigma = 0.01$ S/m.

These properties characterize smooth surfaces. Sometimes, at PCN working frequencies, the ground and walls are rough and a surface roughness parameter, σ_h must be included in the model for each surface.

Information about the roughness parameter and electrical properties of buildings and the ground is not directly available. But many times, one has information relative to the type of building materials and, especially, of the ground, so reflective properties can be indirectly obtained.

3.1.3 Faceted Models

Originally, the environmental data can be in two forms: raster and vectorial [8]. In the first case, the environment is divided into cells and each cell contains the corresponding information. The cell's size depends on the resolution of the scene description. In the vectorial form, the information is associated with geometric entities (lines, polygons, etc.). For the applications of deterministic propagation models, the information must be available in vectorial form. When the original data are in raster form, it is necessary to transform the original information to vectorial form [9].

To incorporate geometric and morphological information into a propagation tool, it is necessary to present such vectorial data in a suitable form. To do this, a database structure must be developed to store and handle the required building and terrain data. In the database, buildings and terrain are stored using a facet model; that is, their surfaces are modeled with polygonal plane facets. Usually, each building wall is represented by a four-sided facet. Roofs are modeled by facets with an arbitrary number of sides. The ground surface is also modeled by plane facets. The number of facets used to model the ground depends on the size of the scene and on the terrain orography. In plane scenarios, a single facet can be enough to model the ground. When information about windows and doors is available, they can be suitably modeled as facets with the adequate morphological properties.

In the database the stored data includes the following:

- Number of facets;
- Number of vertices of each facet;
- Cartesian coordinates of the vertices of each facet;
- Type of material; the type of material is assigned to each facet as an attribute.

As an example, Figure 3.1 shows a simple outdoor scene with one building. The model contains six facets: four for the walls, one for the roof, and one for the ground. Table 3.1 shows how data are arranged in the database. The data in this table provide basic information about the faceted model. Moreover, the databases usually contain other types of data that can be useful for the propagation models:

- Number of buildings;
- Number of facets of each building (from walls and roof);
- Number of facets from walls;
- Number of facets from roofs;
- Number of facets from floors/ceiling;
- Number of facets from ground;
- Type of facet (this indicates if the facet belongs to the ground, a wall, a roof, etc.).

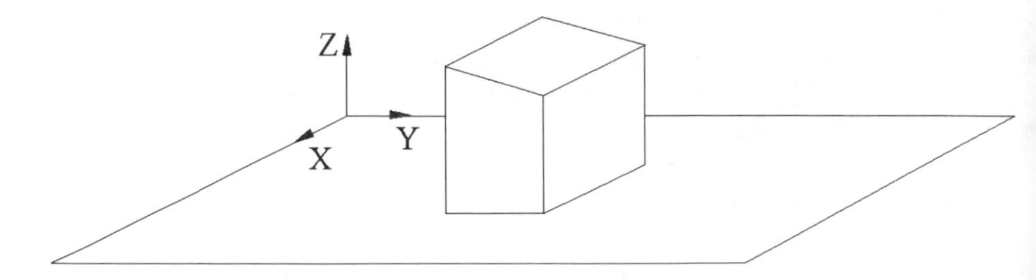

Figure 3.1 A simple 3D scene.

Table 3.1
Data Relative to the Materials and Coordinates of the Facets of Figure 3.1

Facet	No.Ver	Material	X-coords	Y-coords	Z-coords
1	4	Bricks	5.0,5.0,4.0,4.0	7.0,7.0,7.0,7.0	0.0,3.0,3.0,0.0
2	4	Bricks	5.0,5.0,5.0,5.0	7.0,8.0,8.0,7.0	0.0,0.0,2.0,3.0
3	4	Bricks	5.0,4.0,4.0,5.0	8.0,8.0,8.0,8.0	0.0,0.0,2.0,2.0
4	4	Bricks	4.0,4.0,4.0,4.0	8.0,7.0,7.0,8.0	0.0,0.0,3.0,2.0
5	4	Roof Tile	5.0,5.0,4.0,4.0	7.0,8.0,8.0,7.0	3.0,2.0,2.0,3.0
6	4	Asphalt	0.0,8.0,8.0,0.0	0.0,0.0,15.0,15.0	0.0,0.0,0.0,0.0

Usually, it is very useful to store additional geometric data that will be used by the propagation tool repeatedly. This information can be obtained from the description of each facet and usually includes the normal vector of each facet (see Appendix 3A) and the facet's topology. The facet topology information deals with the facets connected to a given facet. For each facet boundary it indicates the connected facet. The number of facets connected to a given facet is in agreement with the number of vertices. If a facet is isolated in any boundary, it appears in the database as connected with the facet number 0.

Table 3.2 shows how the data relative to the model of Figure 3.1 are arranged in the database. This information deals with the facets. But the edges of the model can play an important role in the propagation phenomenon (see Chapter 2). The edges of the model arise from the connection of pairs of facets. They are defined by their endpoints, the pair of facets that forms the edge, and the angle of the wedge (angle formed by the facets), which can be stored as well.

Table 3.3 shows how the edge data of Figure 3.1 are arranged in the database. Edge information can be obtained from the basic faceted model, but

Table 3.2
Data Relative to the Topology and Normal Vectors of the Facets of Figure 3.1

Facet	Topology	Normal Vector
1	2,5,4,6	0.0,–1.0,0.0
2	6,3,5,1	1.0,0.0,0.0
3	6,4,5,2	0.0,1.0,0.0
4	6,1,5,3	–1.0,0.0,0.0
5	2,3,4,1	0.0,0.7071,0.7071
6	0,0,0,0	0.0,0.0,1.0

Table 3.3
Data Relative to the Edges of the Scene of Figure 3.1

Edge	Facets	Angle	Xcoords	Ycoords	Zcoords
1	1,2	90.0	5.0,5.0	7.0,7.0	0.0,3.0
2	2,3	90.0	5.0,5.0	8.0,8.0	0.0,2.0
3	3,4	90.0	4.0,4.0	8.0,8.0	0.0,2.0
4	4,1	90.0	4.0,4.0	7.0,7.0	0.0,3.0
5	1,5	45.0	5.0,4.0	7.0,7.0	3.0,3.0
6	2,5	90.0	5.0,5.0	7.0,8.0	3.0,2.0
7	3,5	135.0	5.0,4.0	8.0,8.0	2.0,2.0
8	4,5	90.0	4.0,4.0	8.0,7.0	2.0,3.0

it is useful to have the edge description available in the database in a straightforward manner. To obtain edge information from a basic faceted model, a basic rule must be followed: A facet boundary cannot be shared by more than two facets.

As mentioned earlier, the database contains three-dimensional information of the environment. Sometimes only 2D data are available from the information sources (plan view maps). In these cases, in outdoor environments, terrain is assumed to be flat, and an approximate height can be chosen for all the buildings in order to obtain a faceted model. Sometimes the number of floors of each building is known. In these cases, approximate building heights can be obtained. Some authors directly work on 2D models in order to facilitate the application of the radio propagation model [10]. In this case, diffractions from rooftops are neglected. This is a valid assumption if building walls are much higher than the transmitter height. Some authors place a limitation such that only reflecting surfaces parallel to the coordinate axis are treated [4].

In the indoor case, 3D faceted models can be obtained by combining a 2D description with information about the height of each floor of the building. Here, a 3D model is indispensable when accounting for the influence of floors and ceilings.

Together with the database, a graphic interface with which to visualize the environment and present output propagation data to users in a suitable way is necessary. Some authors use standard CAD (computer-aided design) packages as a database manager and graphical interface [11, 12]. CAD packages are widely available and often inexpensive.

Most CAD systems support standard graphic formats for describing the models in ASCII files. These formats permit digital exchange of database information among CAD systems and other analysis-type programs that require geometrical information of bodies or scenes (e.g., site-specific radio propagation simulators). The benefit of common formats is that a user does not have to develop special translators for each software system used. The only requirement is a translator to and from the standard format. Today, two standard formats are widely used and accepted by designers and CAD systems vendors: Initial Graphic Exchange Specification (IGES) and Drawing Interchange File (DXF). References [13] and [14] provide complete descriptions of both formats.

Working in this way, CAD programs are used as geographic information systems (GIS). Figure 3.2 shows the connection between the CAD program and the propagation simulator in a propagation prediction system.

Another possibility is to use a classic GIS tool. GIS systems are more suitable for import and management data (especially nongeometric data) and they have more flexible databases [8].

A third alternative is to develop one's own graphical interface and one's own database structure. The advantage of this strategy is that both the graphical interface and the database are oriented to the program simulator; that is, they work exclusively with the necessary data (data of faceted models). On the other hand, the creation of a graphical interface requires the software implementation of visualization algorithms.

3.2 Rationale for Ray Tracing: The Shadowing Problem

The typical urban and indoor scenes tend to be complex environments that require a large number of facets to be modeled (see Figure 3.3). In such environments, radio wave propagation is an extremely complicated phenomenon. In a mobile communication problem, it is necessary to compute the field using a large number of points located along a line in a street (see Figure 3.4),

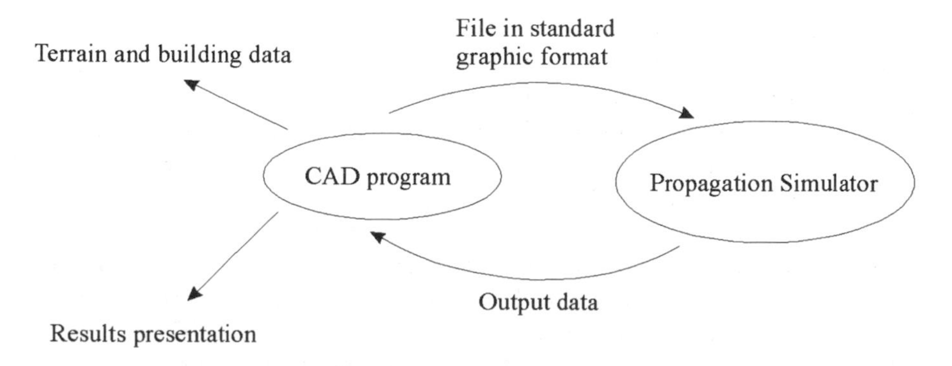

Figure 3.2 Scheme of a propagation system using a CAD program.

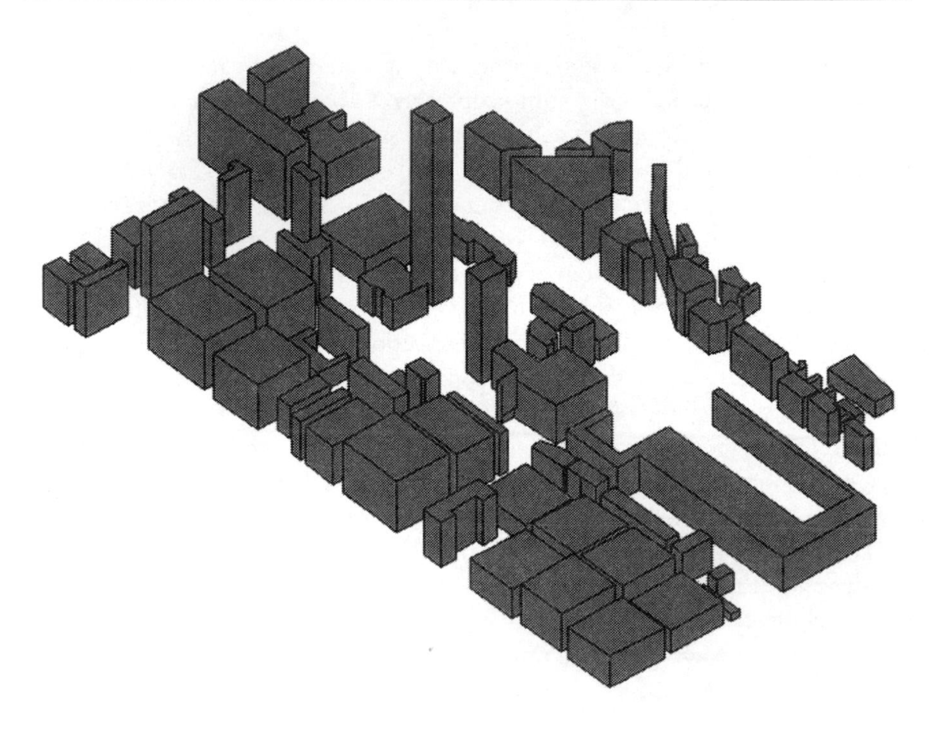

Figure 3.3 Typical 3D urban scenario.

in an indoor corridor, or in the nodes of a mesh (see Figure 3.5). The number of observation points can be in the order of thousands or even greater. The source or sources can be located at any point in the scene.

Even using a simplified model of the scenes as the faceted model, and a ray propagation approach such as the GTD/UTD, the problem of simulating the propagation of radio waves is a complex task. Figure 3.3 shows a faceted model of a typical urban scene.

In the computer graphics world, "shadowing test" is known as the operation to determine if an observation point is visible from a source point. In other words, the shadowing test determines if any object of the scene occludes a given ray path from a source point to an observer point. The source points can be the transmitter antennas, reflection points, diffraction points, and transmission points. The observers can be observation points, reflection points, diffraction points, or transmission points.

In faceted models, the environmental objects are described with facets, so the shadowing test is reduced to repeatedly applied ray-facet intersection tests. Let's consider a faceted model of an urban environment with N_f facets, N_e edges, a transmitter antenna and N_0 observation points. Using a "brute force" method, the ray-facet intersection test is performed a number of times proportional to:

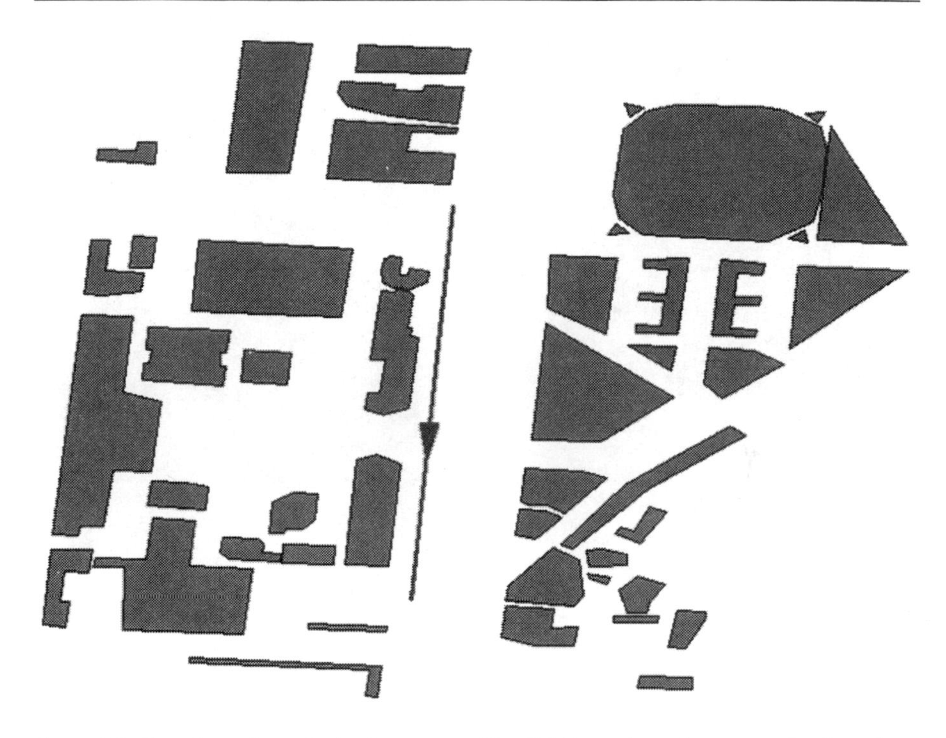

Figure 3.4 Top view of a typical urban scene. The field is evaluated as a set of points along the line.

- $N_0 N_f(N_f + N_e)$ for first-order effects (direct rays and simple-reflected rays and simple-diffracted rays);
- $N_0 N_f(N_f + N_e)^2$ for second-order effects (double-reflected rays, reflected-diffracted rays, diffracted-reflected rays, etc.);
- $N_0 N_f(N_f + N_e)^3$ for third-order effects (triple-reflected rays, diffracted-reflected-reflected rays, etc.) and so on.

A typical microcell can be modeled by a number of facets and edges of the order of several hundreds, so, using only a "brute force" method, the number of tests required for the ray tracing can be incommensurate. In such cases, it is necessary to reduce, as much as possible, the number of ray-facet intersection tests using efficient ray propagation models and ray-tracing acceleration techniques. These techniques are shown in the scheme of Figure 3.6. They will be addressed, in detail, throughout Sections 3.3 and 3.4.

Even using such algorithms in complex environments, the time consumed in ray-facet intersection tests can take up more than 90% of the total ray-tracing time. The remaining 10% is wasted in the computation of reflection points in facets, diffraction points in edges (Section 3.5), etc.

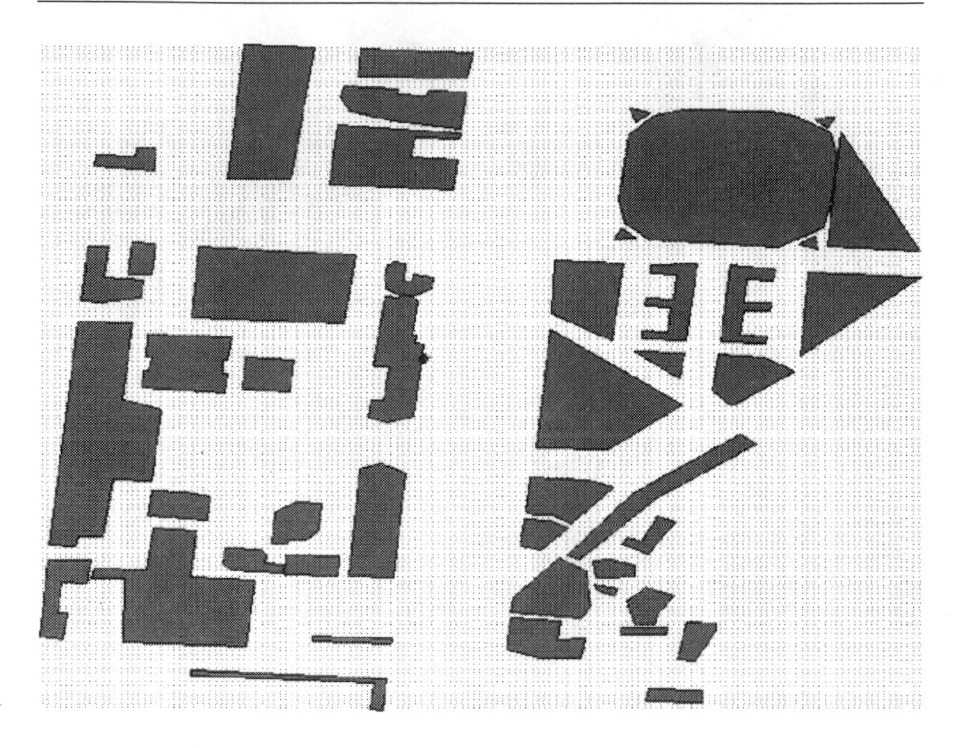

Figure 3.5 Top view of a typical urban scenario. The field is evaluated in a mesh of points.

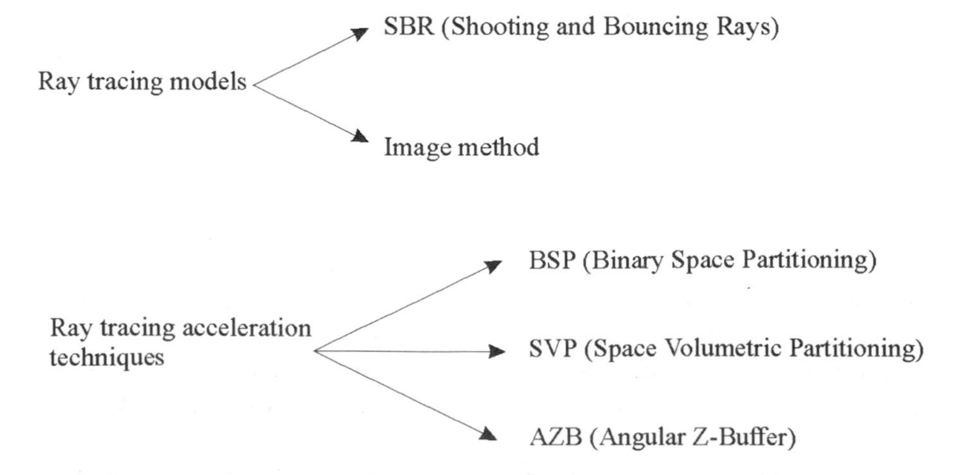

Figure 3.6 Ray-tracing models and ray-tracing acceleration techniques.

3.3 Ray Propagation Models

Ray tracing is an important topic in the world of computer graphics. Several ray propagation models have been developed for computer visualization applications [15,16]. They simulate the propagation of light, that is, electromagnetic waves in the visible frequency range. These models assume some simplifications valid in this frequency range. They are based on a scalar formulation that does not take into account the polarization and phase of the fields associated with the rays. Moreover, the edge diffraction phenomenon is not taken into account because it is of little importance. Consequently, the ray-obstacle interactions are reduced to reflections and transmissions.

The assumptions just given are not valid for lower frequencies such as the mobile communication range (UHF), so no ray-tracing models for computer visual applications can be directly applied to the UHF problem. At these frequencies, propagation models require a vector formulation that considers the vectorial nature, including phase and polarization, of the individual contributions to the total field. Moreover, apart from reflections and transmissions, additional phenomena such as edge diffractions can play an important role, so they must be included in the model.

To show the importance of edge diffractions in the UHF range, simulations with and without diffracted rays have been considered for a typical urban scene. In particular, the following cases are presented here:

- Case 1: Only direct and reflected rays in Figure 3.7.
- Case 2: Direct, reflected, and diffracted rays in Figure 3.8.
- Case 3: All the rays of case 2 plus double-reflected rays, diffracted-reflected rays, and reflected-diffracted rays in Figure 3.9.

In all of these cases, the transmitter antenna is omnidirectional. It is 5.5m high. The frequency is 922.2 MHz. The observation points are in the same plane parallel to the ground and 1.5m high.

Examining Figures 3.7, 3.8, and 3.9, one can conclude that wide coverage is achieved by the diffraction mechanism, which becomes the primary factor in the shadow areas for direct and reflected rays. In some works [11, 12, 15] the coverage in these areas is computed considering reflections of very high order. But using this strategy, the ray-tracing is more complicated and less efficient under a computational point of view. Moreover, in general, the diffracted field provides more accurate field values than the high-order reflections [17, 18]. An exception must be made for long indoor corridors where the contribution of high-order reflections is important.

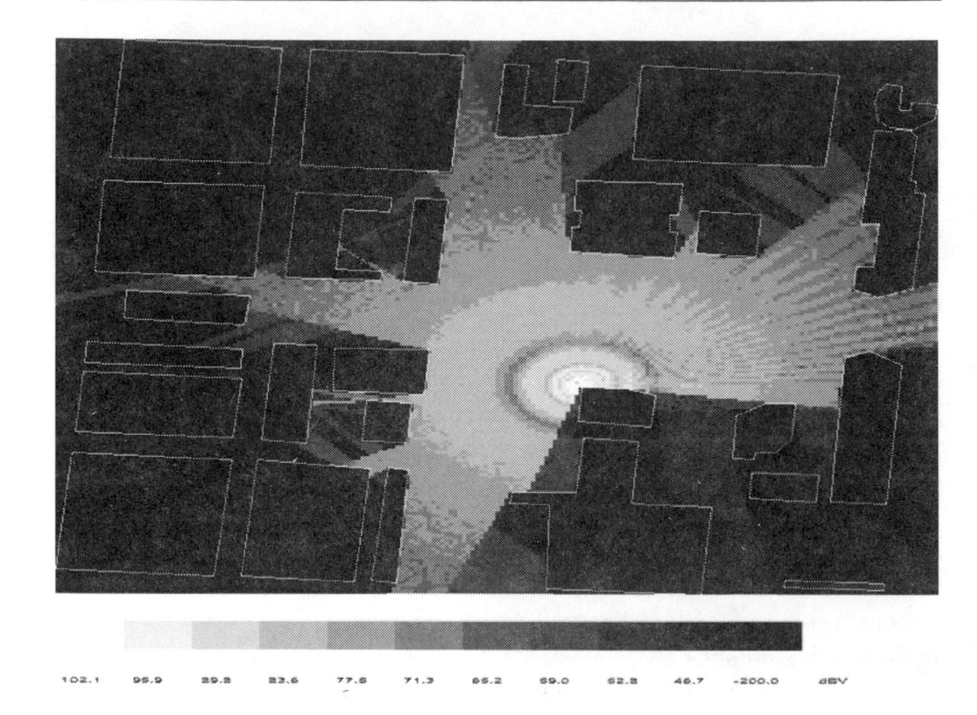

Figure 3.7 Field coverage level by the direct and singly reflected rays.

Due to their common aspects, most of the propagation models in mobile communications are inspired by or incorporate techniques of visual rendering that have been adapted to the special characteristics of the UHF wave propagation. The strategies proposed for the propagation prediction in the UHF range can be classified into two main groups: techniques based on the shooting and bouncing rays method [11, 12, 19] and techniques based on the image method [4, 10, 20].

3.3.1 The Shooting and Bouncing Rays Method

The shooting and bouncing rays (SBR) method is also called the pin cushion or ray-launching method. The SBR method has been widely used in radar cross section (RCS) and scattering problems [21–25]. This is a forward (or direct) ray-tracing model because the propagation simulation is performed from the source following the rays in their propagation. Therefore, the real process of radio wave propagation is modeled.

In the SBR method, ray tubes are shot from the source (transmitter antenna) covering all directions in space. Each ray tube occupies the same solid angle. When the tube advances, its cross section increases; in other words, the

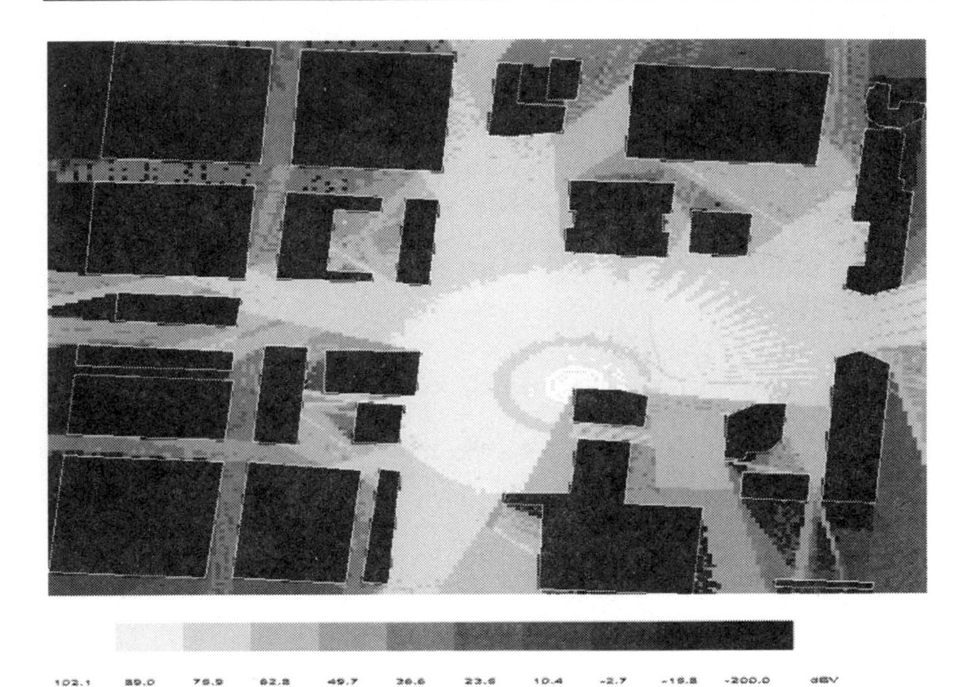

Figure 3.8 Field coverage level by direct, singly reflected, and singly diffracted rays.

rays diverge. To keep the angular distance between adjacent ray tubes constant (or nearly constant), a method based on the theory of geodesic domes is used. This technique is presented in detail in [12]. Obviously, when the number of rays launched from the source increases, the angular separation of rays decreases. As an example, for a maximum angular separation of one degree between rays, 40,000 source rays must be traced in all directions. Using this technique, one obtains tubes of rays that completely cover the space surrounding the source. Moreover, there are no intersections between the ray tubes, that is, two ray tubes do not share any part of the space.

As they propagate, the rays encounter environmental obstacles. In this case, reflection occurs and the propagation continues in another direction. Notice that for each ray, the shadowing test must be performed on every obstacle (facet) of the scene. This is a computationally intensive process that requires ray-tracing acceleration algorithms. The propagation of each ray is followed until its field intensity level drops below a certain threshold level. In this case the algorithm "forgets" the ray. Selecting a threshold is similar to having a receiver noise floor below which nothing can be received. The threshold level must be carefully chosen. If it is low, the algorithm wastes CPU and memory size without a significant improvement in the results. On the other

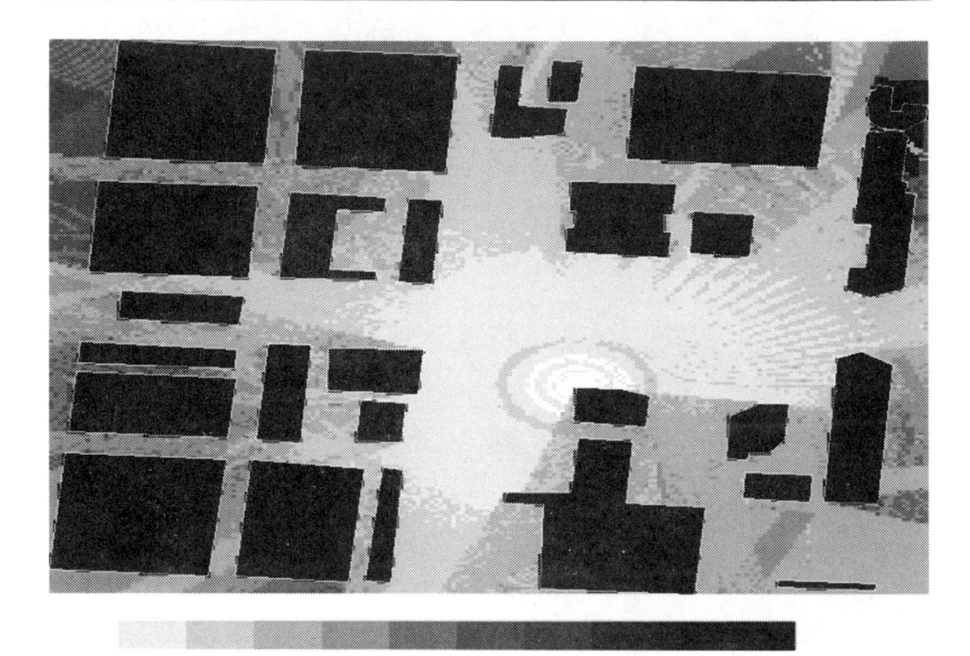

Figure 3.9 Field coverage level including direct, reflected, diffracted, double-reflected, diffracted-reflected, and reflected-diffracted rays.

hand, if the threshold level is too high, important contributions to the field in the receiver antenna can be lost.

In a real urban or indoor environment, the number of ray traces tends to be very high. The whole ray propagation process can be arranged and stored in the so-called "ray-tracing tree" [12]. The tree branches are the rays emitted from the transmitter and the tree knots are the obstacles (facets) of the environment encountered by the rays in their propagation. As Snell's law states, when a ray reaches an obstacle, it is transmitted and reflected so each ray is decomposed into two "children" when it reaches the obstacle. In other words, the ray-tracing tree is a binary one. But in outdoor propagation, the transmitted rays tend to be discarded so only the branch corresponding to the reflected ray is taken into account. In each knot, the parameters required by the GO to compute the reflection coefficients (angle of incidence, length traveled by the ray, type of material of the facet, etc.) are stored. The tree contains all the information about the ray trace in a suitable format. Notice that the ray-tracing tree is independent of the observer's location. Figure 3.10 shows a ray-tracing tree for a simple 2D outdoor scene. For the sake of simplicity, only four rays have been traced. In practice, thousands of rays are launched from the transmitter antenna.

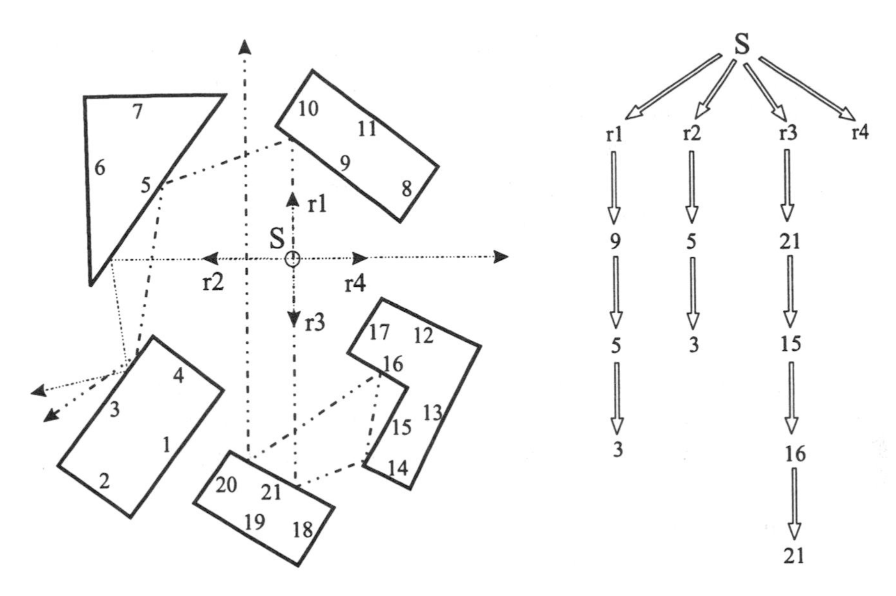

Figure 3.10 Example of a ray-tracing tree for a simple 2D outdoor scene.

For a given observer location, using the tree information, the rays that reach the observer are determined and then the field strength is calculated. As stated before, the transmitter and observer are modeled as point sources, and the number of rays traced from the transmitter is finite. Therefore, in order to determine if a ray reaches the receiver, a reception sphere [24] centered in the receiver point is considered. When a ray intersects the sphere, it is considered received and it contributes to the total received signal at this point. An adequate value for the radius of the reception sphere (R) is

$$R = \frac{\alpha d}{\sqrt{3}} \tag{3.1}$$

where α is the angular separation between the rays launched from the source. The parameter d is obtained by computing the total (unfolded) path length (d) of the ray from the transmitter to the projection point [11,12]. This last point is the perpendicular projection of the observation point on the ray path. The reception sphere effectively accounts for the divergence of the rays from the source and ensures the uniqueness of all of the reflection points. For typically sized microcells and values of α of less than one degree, the radius of the reception spheres will be of the order of meters.

Using the receiver sphere technique, the magnitude of the field (direct or reflected) in the receiver point is well approximated. But the phase of the field can differ considerably with respect to the phase of the exact ray that passes exactly throughout the observation point. Consequently, unlike other propagation models, the total field strength at the receiver cannot be obtained from a coherent sum of the individual contributions. However, it is possible to calculate the mean field strength ($\overline{E}_R$) from the magnitudes of the individual multipath components (E_i) as follows:

$$\overline{E}_R = \sqrt{\Sigma E_i^2} \tag{3.2}$$

The SBR technique presents serious difficulties when the edge diffraction contribution is incorporated into the model. When a ray reaches an edge of a wedge, it is diffracted. As Keller's law states, the resulting diffracted rays (infinite) propagate in all directions on the Keller's cone (see Chapter 2). Therefore, an edge incident ray produces infinite diffracted rays, so it is impossible to incorporate the edge-diffracted rays in a forward model.

3.3.2 Image Method

The image method for ray-tracing is based on the image theory (see Chapter 2). It provides an efficient technique to compute the rays' reflections in environments modeled with plane facets.

Given a source point (S) and a facet, the reflected rays in the facet can be considered as rays radiated directly from a virtual source point, called the image source (I). The image source is located symmetrically to S with respect to the plane that contains the facet (see Figure 3.11). Notice that the position of I depends exclusively on the location of S and on the position and orientation

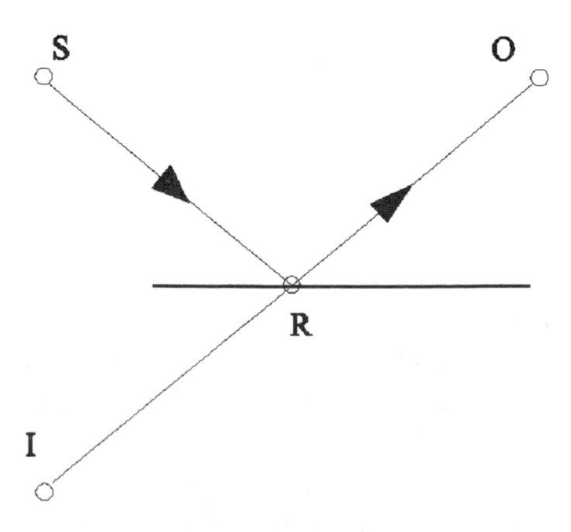

Figure 3.11 Application of the image method to the ray reflections.

of the facet. Therefore, I is independent on the observation point. The field radiated by the image source is obtained from the radiation characteristics of the real source (S) and the electric properties of the facet (see Chapter 2).

For a given observation point (O), the reflection point (R) is easily calculated as the intersection between the segment I-O and the facet. Section 3.6 deals with efficient ray-facet intersection algorithms.

In a scenario modeled with N flat facets, the number of images will be N. Consequently, the maximum number of reflected rays that reach an observation point is N. Obviously, in the real world, the number of reflected rays reaching an observer is lower. This is because of two reasons:

1. Due to the finite dimensions of the facets, only observers located on the reflection space (RS) (see Figure 3.12) of the facets can receive reflected rays. In practice, an observer is in the RS of a facet when R is inside the facet. Section 3.6 deals with efficient algorithms to determine if a point is inside a facet.
2. The second reason is that the reflected ray or the incident ray (from S to R) can be hidden by other facets of the environment. It must be achieved by the shadowing test.

Double-reflected rays can be analyzed in a similar way. The sources of double-reflected rays are the images of the first-order images (images of the single reflections). They are called second-order images. The number of second order images will be $N(N-1)$. Three conditions must be satisfied for a double reflection:

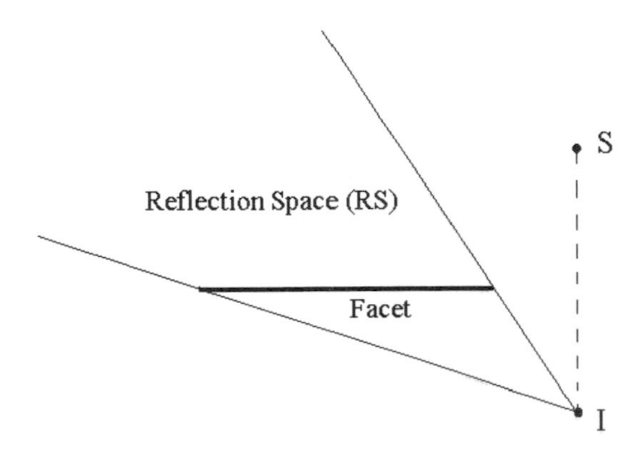

Figure 3.12 The reflection space (RS) of a reflecting facet.

1. The observation point must be in the RS of the second facet. In other words, the second reflection point must lie in the second facet.
2. The second reflection point must be in the RS of the first facet. In other words, the first reflection point must lie in the first facet.
3. None of the three paths shown in Figure 3.13 ($S–R_1$, $R_1–R_2$, $R_2–O$) can be hidden by other facets of the environment (shadowing tests).

The multiple reflections can be analyzed in the same way. The number of images for the *k*-order reflections will be

$$N(N-1)^{k-1} \tag{3.3}$$

The images can be arranged in a tree graph called the *images tree*. Initially, the first branching is N-fold; and the later ones are $(N-1)$-fold. The image method ensures that all the rays that suffer a given number of reflections (or fewer) are tracked because all the potential reflected rays are stored in the images tree.

It is important to note that using the image method, one only manages a restricted number of rays, avoiding the treatment of many irrelevant rays that leave the source but do not reach the observers in less than a number of reflections.

Given a source (S) (it could be the real source or an image source), when a facet (F) is totally hidden by another (F_1), its image (I) can be removed from the images tree and all of the images obtained from I can also be removed (see Figure 3.14). This simplifies the images tree and, consequently, produces an important savings in memory and CPU time. Algorithms for analyzed facets hiding can be found in [25].

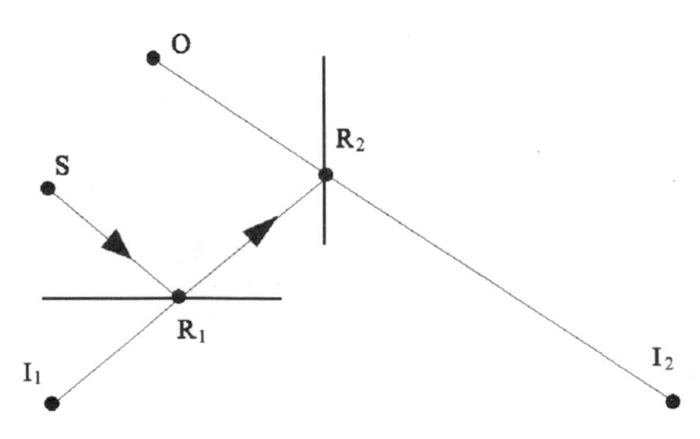

Figure 3.13 Application of the image method to double reflections.

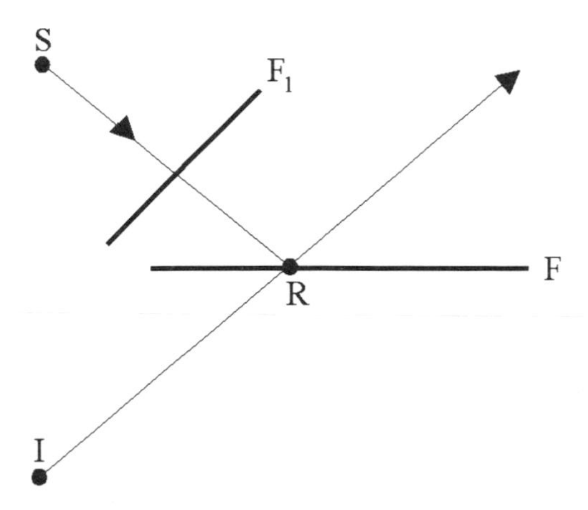

Figure 3.14 Image discarded by hiding.

The tree can also be simplified considering the facets' location with respect to the RSs. If a facet F_1 is totally outside of the RS of a facet F, there is no double reflection S–F–F_1 (see Figure 3.15). Therefore, the image of F_1 that hangs on to the image of F can be removed from the tree. Obviously, subsequent images in the tree can also be removed.

Moreover, in outdoor scenarios the backface culling algorithm (see Appendix 3B) must be used to discard images in the tree. Figure 3.16 shows an example of an images tree for a 2D outdoor environment. The tree has been simplified by applying the backface culling algorithm and discarding facets

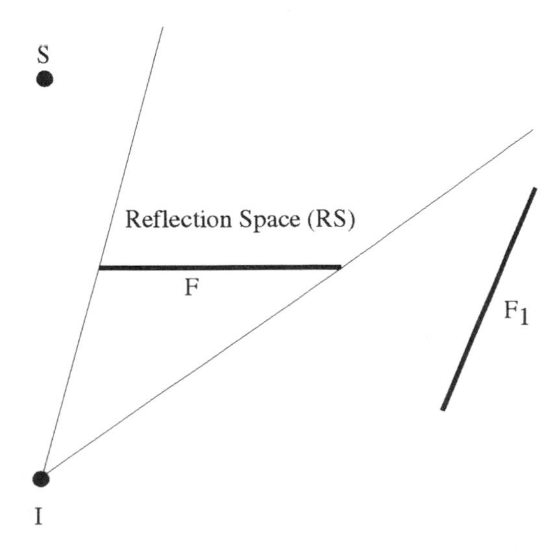

Figure 3.15 Facet F_1 is outside the RS of facet F so it cannot be a double reflection S–F–F_1.

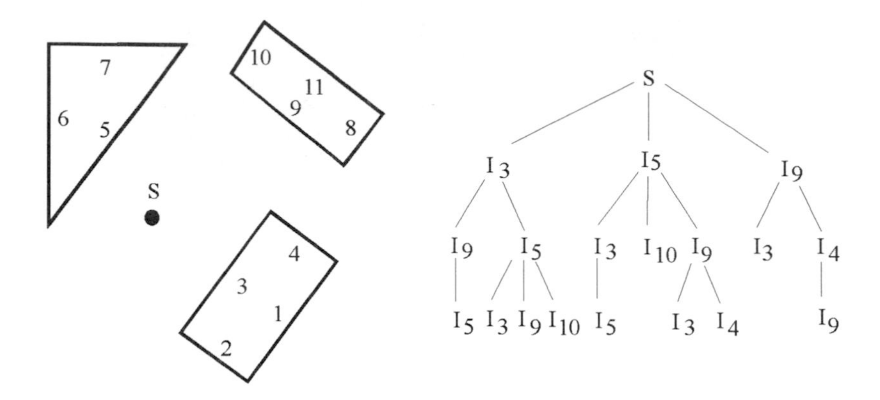

Figure 3.16 Example of an images tree of a simple 2D outdoor scene.

outside the corresponding RS. Only three levels of reflection have been considered.

The interrogation of the images tree is achieved in the backward sense. When one is interested in the k-order reflected rays that reach a given observation point, the analysis must start on the knots of the k-level (k-order images) of the tree. For each of them, the procedure is as follows.

The last reflection point (R_k) is calculated and we determine if it lies in the corresponding facet. In other words, we determine whether the observer is in the SR of the facet. If the answer is negative, there is no reflection and the algorithm continues with another knot of the tree. Otherwise, a check is made to see if the path R_k–O is hidden by another facet of the scene (shadowing test). If it is not hidden, the previous reflection point is calculated and we determine whether it lies in the corresponding facet. In other words, a check is made to see if R_k is in the SR of the previous image (I_{k-1}). If the answer is positive, the shadowing test for the path R_{k-1}–R_k is achieved. If the test is negative, the algorithm continues calculating the previous reflection point and so on. The algorithm finishes when one arrives at the real source.

When the order of the reflections to consider and the number of facets increase, the number of shadowing tests increases meaningfully. Therefore, the image method should be combined with ray-tracing accelerating algorithms that permit one to reduce the number of shadowing tests. These algorithms are presented in Section 3.6.

Image theory permits one to obtain the exact path followed by the reflected rays, so it is possible to obtain the phases of the fields associated with the reflected rays. Therefore, the multipath contributions at the observation points can be summed coherently to obtain the received field.

3.4 Ray-Tracing Acceleration Techniques

As stated in previous sections, when the environments are complex, the ray-tracing process should be accomplished with algorithms that will improve the efficiency of the simulations. Therefore, ray-tracing acceleration algorithms are a common topic in the world of ray-tracing [15, 16]. Some of these techniques can be applied to electromagnetic wave propagation in the UHF band.

The ray-tracing acceleration techniques can be classified into four categories according to their goal [15]:

1. *Reducing the cost of intersecting a ray with the primitives used in the model.* The facet models use only the polygonal plane facet as primitive. Geometrically, the facet is a very simple primitive so the cost of the intersection ray-facet tests is reduced. The ray-facet intersection algorithms are a well-known topic (see Section 3.6).
2. *Reducing the total number of ray-primitive (ray-facet) intersection tests.* In all the propagation models presented, the shadowing test must be achieved repeatedly. Therefore, the number of ray-facet intersection tests carried out during the ray-tracing simulation tends to be extraordinarily large. Although the ray-facet intersection test is a computationally cheap process, the great number of tests in urban environments requires large CPU times to accomplish the simulations. The efficiency of the shadowing tests can be dramatically improved by discarding facets and, therefore, reducing the number of ray-facet intersection tests. These accelerating techniques are the most efficient in complex faceted models. The accelerating algorithms presented in the following sections belong to this category.
3. *Reducing the total number of rays intersected with the environments.* Sometimes, during the ray-tracing process, it is possible to estimate the contribution of a ray in the final result. In this case, if the contribution is below a certain threshold level, the ray can be discarded. With this technique the number of rays to be traced can diminish considerably. This technique is especially suitable in SBR-based models as was mentioned in Section 3.3.1.
4. *Replacing individual rays with a more general entity.* This category contains a number of techniques that begin by replacing the familiar concept of a ray with a more general entity (generalized ray) that subsumes rays as a special (degenerate) case. For instance, cones of both circular and polygonal cross sections have been used as generalized

rays. The basic idea of these methods is to trace many rays simultaneously. Although this method has been used successfully in light propagation problems, it seems difficult to make it compatible with the GTD.

The algorithms proposed in this book are inspired by techniques used in computer graphics. They belong to the second category, so their objective is to reduce the number of ray-facet intersection tests. They are easy to use in combination with the UHF propagation models presented in Sections 3.3.1 and 3.3.2. They are the binary space partitioning (BSP), the space volumetric partitioning (SVP), and the angular *z*-buffer (AZB). The AZB is suitable for propagation models based on the image method, whereas the BSP and the SVP are appropriate for both propagation strategies—the SBR and the image method.

Apart from those, backface culling is another basic and simple technique that can be used to accelerate the ray-tracing that must always be used (see Appendix 3B).

3.4.1 The Binary Space Partitioning Algorithm

The BSP algorithm [27] is an efficient method for calculating and storing the visibility relationships among a group of facets in a 3D space. This information is stored in a binary tree structure called the BSP tree.

As for many of the ray-tracing acceleration techniques, the main objective is to reduce the number of ray-facet intersection analyses in the ray-tracing process. It is achieved by using the information from the BSP tree. By reducing the number of ray-facet intersection questions, the time consumed by the ray-tracing procedure can be diminished dramatically.

In a ray-tracing model based on the BSP, the first task is to generate a BSP tree of the scene. It depends exclusively on the geometry of the scene so it is independent of the sources and observer's locations. Afterwards, during the ray-tracing analysis, the BSP tree is interrogated repeatedly in order to reduce the number of ray-facet intersection questions.

The BSP tree is a binary tree structure that contains information about the relative positions of the facets in a 3D (or 2D) scene. Each tree knot represents a facet. The BSP tree's root is any facet of the scene; the algorithm works correctly regardless of which one is selected. The root polygon is used to partition the environment into two half-spaces. One half-space contains all remaining polygons in front of the root polygon, relative to its normal surface (see Appendix 3A); the other contains all polygons behind the root polygon. Any polygon lying on both sides of the root polygon's plane is split by the plane and its front and back pieces are assigned to the appropriate half-space. One polygon each from the root polygon's front and back half-space becomes its front and back children, and each child is recursively used to divide the remaining polygons in its half-space in the same fashion. The algorithm terminates when each node contains only a single polygon.

For example, for the 2D outdoor scene of Figure 3.17, the tree has been generated as follows: Facet 3 has been chosen as the root knot. The plane that contains facet 3 divides the space into two half-spaces that correspond to the branches of the knot. The half-space over the facet (pointed out by the normal vector of the facet) corresponds to the right branch, and the half-space under the facet corresponds to the left branch. The half-plane splits facets 5 and 4, so two new facets for each one are assigned to the appropriate half-plane. One of the facets arbitrarily chosen over the plane is placed in the right branch (facet 6), and a facet placed under the plane is hung from the left branch (facet 1). The process is then repeated from the "children" facets (1 and 6). Following the above procedure in a recursive way, the binary tree is created. Figures 3.18 through 3.21 illustrate the tree generation for the present example.

The tree contains the information about the relative positions of the facets and it is independent of the source and observer locations. For a given scene, many trees can be created, but not all are equally efficient as will be shown later. Figure 3.22 shows an alternative tree for the previous scene. The BSP tree information is used to reduce the number of ray-facet intersection questions in the ray-tracing process.

Depending on the effect analyzed and on the ray path considered, the source can be a transmitter antenna, a point of reflection, a point of diffraction, or a point of transmission. The observer can be an observation point, a point of reflection, a point of diffraction, or a point of transmission.

Given a source (S) and an observation point (O), the root knot is initially considered. There are three possible locations of S and O with respect to the root facet:

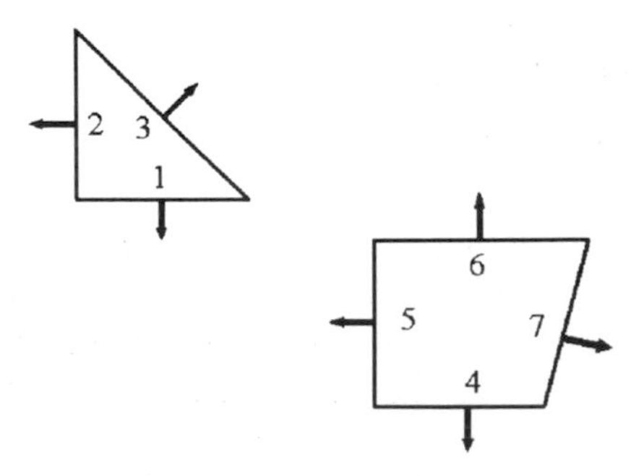

Figure 3.17 Two-dimensional outdoor scene analyzed.

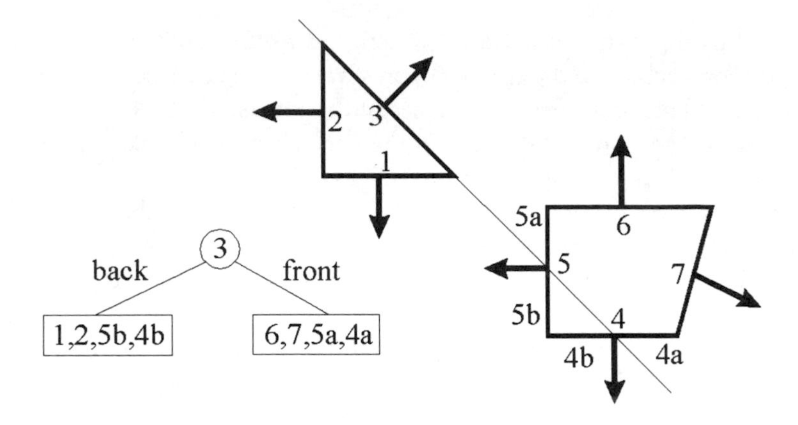

Figure 3.18 BSP tree generation (I).

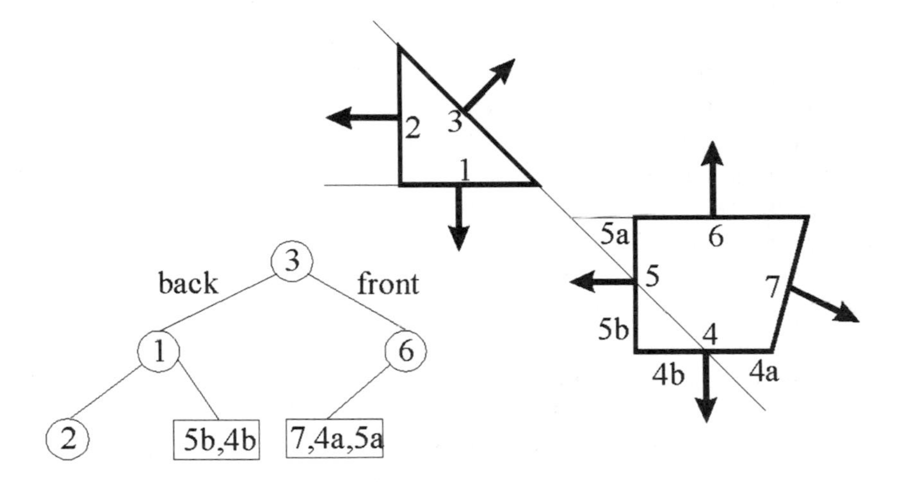

Figure 3.19 BSP tree generation (II).

1. S and O over the facet;
2. S and O under the facet; or
3. S and O in different half-spaces.

In the first case, the present facet plus all those suspended from the left branch cannot hide the path S–O. So one goes to the facet hung on the right branch and the procedure continues with the remaining part of the tree. Notice that, by analyzing one facet, all the facets located in one half-space have been rejected.

In the second case, the present facet plus all those hung from the right branch cannot hide the path S–O. Then, one proceeds to analyze the facet

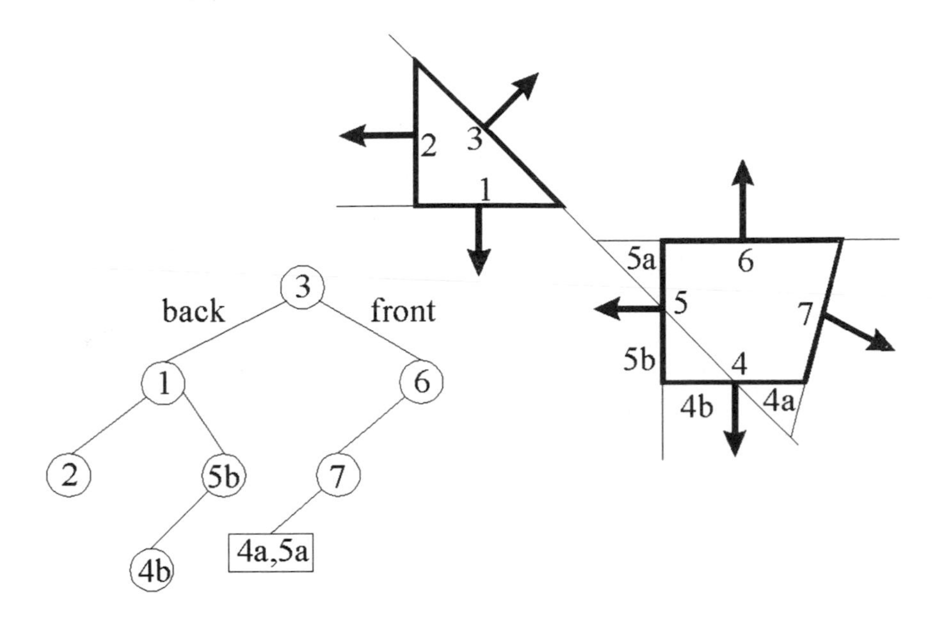

Figure 3.20 BSP tree generation (III).

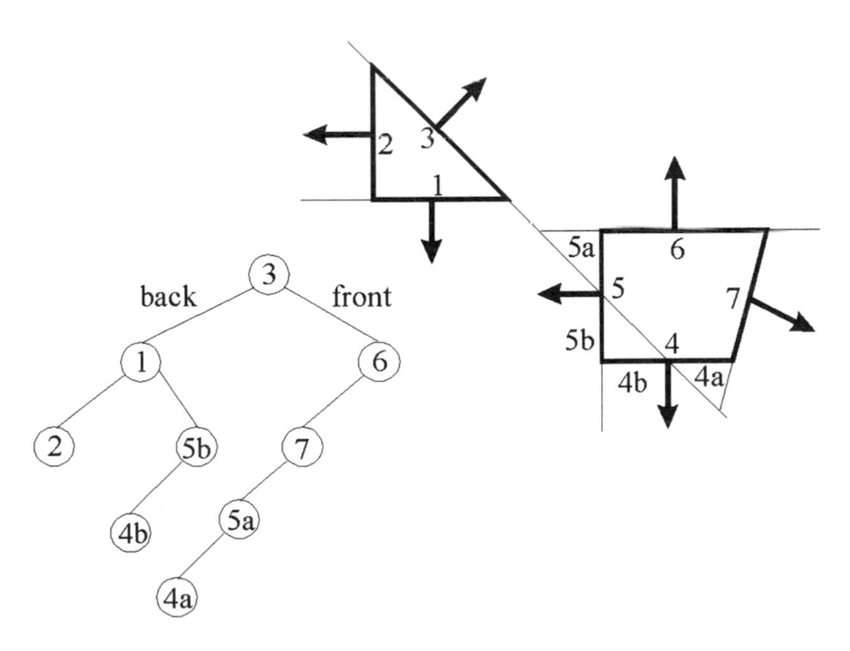

Figure 3.21 BSP tree generation (IV).

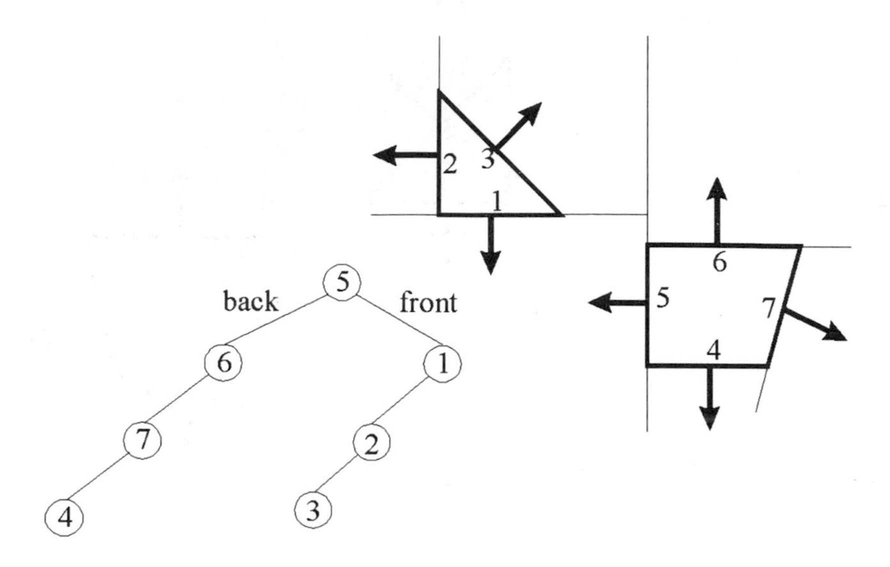

Figure 3.22 Another example of a BSP tree.

suspended from the left branch and the procedure continues with the remainder of the tree.

In case three, the facet can hide the path S–O and a rigorous test must be made. If it does not, the procedure continues with the children facets.

Apart from the ray-facet intersection tests, the unique operation achieved during the tree interrogation is to determine if a given point P (it can be the source point or the observer point) is over or under a facet. This test is reduced to analyze the sign of the scalar product between the normal vector $\hat{n}$ and the vector $\vec{v}$, which joins a facet vertex with P (see Figure 3.23).

If $\vec{v} \cdot \hat{n} > 0$, P is over the facet (see Figure 3.23(a)).
If $\vec{v} \cdot \hat{n} < 0$, P is under the facet (see Figure 3.23(b)).

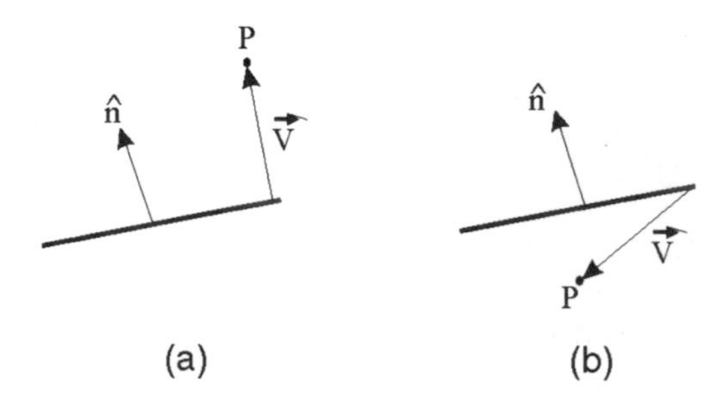

Figure 3.23 Point P over and under the facet. Cases (a) and (b), respectively.

The underlying idea of the BSP algorithm is to interchange the computationally expensive ray facet test with the above scalar product. In complex scenes this provides a very important reduction in the ray-tracing computation time with respect to the brute force method.

As already mentioned, various trees can be created from a given scene. The efficiency of the algorithm depends a lot on the tree structure. The optimum trees must satisfy two conditions:

1. The number of split facets is low. In these cases, the number of knots is close to the number of facets in the scene.
2. The tree is balanced; that is, it has only a few depth levels.

The minimum number of depth levels is $\text{INT}(\log_2 N) + 1$, N being the number of facets. In optimum trees, the number of depth levels is close to the above value. For instance, in the tree of Figure 3.21, two facets have been split and the number of depth levels is 5. For the tree of Figure 3.22, no facets have been split and the number of levels is 4. Consequently, the second tree is better. In this scene (seven facets), the ideal tree would have $\text{INT}(\log_2 N) + 1 = 3$ levels.

For an ideal tree in the shadowing test, the maximum number of ray-facet intersection tests (the worst test) will be $\text{INT}(\log_2 N) + 1$. In the best case, no intersection tests are required. The application of the BSP in a ray-based radio propagation model is straightforward:

- In the case of the direct ray, S is the transmitter antenna and O is the receiver antenna. There is a unique path to analyze, so for each observer the tree is interrogated once.
- In the case of reflected rays, there are two paths to consider. First, from the transmitter antenna (S) to the reflection point (O) and, second, from the reflection point (S) to the receiver antenna (O). Therefore, the BSP tree can be interrogated once (if the first path is hidden, the interrogation stops) or twice.
- In diffracted rays, there are also two paths to consider. First, from the transmitter antenna (S) to the diffraction point (O) and, second, from the diffraction point (S) to the receiver antenna (O). Therefore, the tree is interrogated once or twice.
- In second-order effects (double reflections, double diffractions, reflections-diffractions and diffractions-reflections), there are three paths to be analyzed, so the BSP tree is interrogated once, twice, or three times. For the first path, S is always the transmitter antenna and O can be

a reflection or diffraction point. The second path goes from a reflected or diffracted (S) point to another reflected or diffracted point (O). And, finally, the third path comes from a reflection or diffraction point (S) and arrives to the receiver antenna (O).

In the previous 2D scene, Figure 3.24 shows a transmitter antenna (Tx) and four different locations for a receiver antenna (Rx_1 to Rx_4). For the different effects, Table 3.4 shows the number of rays that arrive at the receivers (first number in the table), and the number of ray-facet tests required using the BSP tree of Figure 3.21 (second number in the table). In the application of the brute force method, all facets are tested. The backface culling algorithm (see Appendix 3B) was not applied in any case. Table 3.5 shows the results obtained for the tree of Figure 3.22.

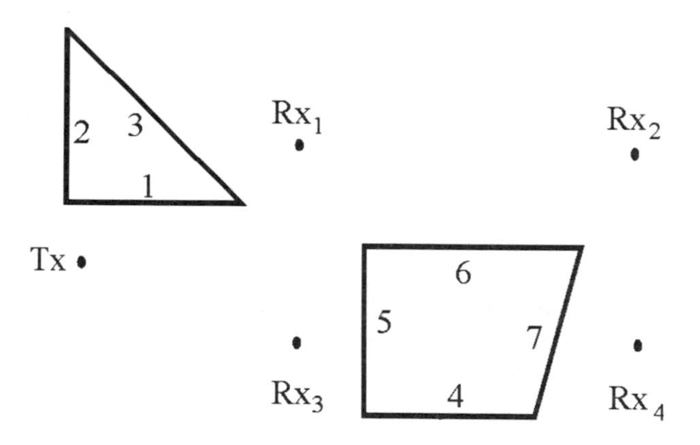

Figure 3.24 Transmitter and receiver locations in the 2D scene.

Table 3.4
Results for the Tree in Figure 3.21

Electromagnetic Effect	Rx_1	Rx_2	Rx_3	Rx_4
Direct rays	0, 1	1, 6	1, 0	0, 3
Reflected rays	0, 0	0, 0	2, 1	0, 0
Diffracted rays	3, 4	2, 1	4, 2	0, 0
Double reflected rays	0, 0	0, 0	1, 3	0, 0
Reflected-diffracted rays	2, 3	1, 0	2, 1	0, 0
Diffracted-reflected rays	0, 0	0, 0	2, 5	0, 0
Double diffracted rays	9, 11	7, 5	8, 6	2, 2

Table 3.5
Results for the Tree in Figure 3.22

Electromagnetic Effect	Rx_1	Rx_2	Rx_3	Rx_4
Direct rays	0, 1	1, 5	1, 0	0, 1
Reflected rays	0, 0	0, 0	2, 0	0, 0
Diffracted rays	3, 3	2, 1	4, 0	0, 0
Double reflected rays	0, 0	0, 0	1, 0	0, 0
Reflected-diffracted rays	2, 3	1, 0	2, 0	0, 0
Diffracted-reflected rays	0, 0	0, 0	2, 0	0, 0
Double diffracted rays	9, 13	7, 6	8, 0	2, 2

3.4.2 The Space Volumetric Partitioning Algorithm

One of the earlier techniques to reduce the number of ray-facet intersection tests is the SVP method. This is also called the uniform spatial subdivision algorithm. Using this technique, the 3D space surrounding the environment is divided into voxels. Voxels are cubes with sides parallel to the coordinate axis. All the voxels together constitute a volume that contains the environment. When the size and shape of the voxels are equal, the space subdivision is said to be *uniform.* Hereafter, a uniform space division is assumed.

For each voxel, the facets that lie totally or partially inside are determined. This information is stored in the SVP matrix, which will be interrogated repeatedly in the shadowing tests. Notice that the SVP matrix is independent of the transmitter antenna and observer's locations; in other words, it depends exclusively on the scene. As an example, Figure 3.25 shows a simple 2D outdoor environment and the facet storage in the voxels.

When a shadowing test for a ray path source-observer is achieved, the voxels pierced by the ray are determined. This can be done very efficiently by incremental calculation [28] when the spatial partitioning is uniform, that is, when all the voxels are equal in size and shape. The only facets that must be tested for intersection are those stored in the voxels pierced by the ray. This can potentially eliminate a vast majority of the facets in the environment from consideration.

When one uses an SBR-based propagation model, it is necessary to determine the closest facet pierced by the rays. In such cases, an important observation is that the rays impose a strict ordering on the pierced voxels from the voxel that contains the source to the voxel of the observer. This ordering guarantees that all intersections occurring in one voxel are closer to the ray origin than those in all subsequent voxels. Therefore, by processing the voxels

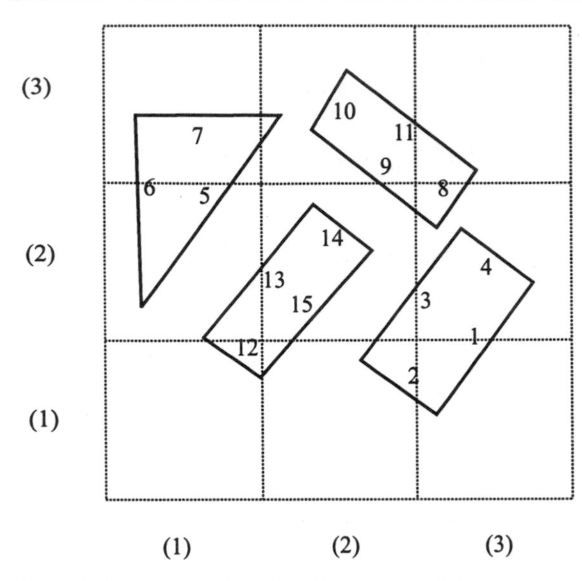

voxel (1,1): 12
voxel (1,2): 5,6,12,13
voxel (1,3): 5,6,7
voxel (2,1): 2,3,12,15
voxel (2,2): 3,13,14,15
voxel (2,3): 5,7,9,10,11
voxel (3,1): 1,2
voxel (3,2): 1,3,4,8,9
voxel (3,3): 8,11

Figure 3.25 Space subdivision in voxels and facets storage for a simple 2D outdoor scene.

in the order in which they are encountered along the ray, it is not necessary to process the subsequent voxels once a ray-facet intersection has been found.

The size of the voxels is important to the efficiency of the SVP algorithm. When the size of the voxels increases, the number of discarded facets decreases. On the other hand, when the size of the voxels decreases, the number of voxels increases, so the number of voxels to process for each ray increases. Moreover, the memory requirements to store the scene increases with the number of voxels. From the author's experience, voxel dimensions about the size of the facets is a good choice.

The SVP technique can be improved using hierarchically partitioned structures [29] or the octree technique [30].

3.4.3 The Angular Z-Buffer Algorithm

The AZB technique has been developed by the authors. It is similar to the light buffer technique [31], although the AZB has a lot of particular features that make it especially well suited for the UHF propagation problem, especially for the treatment of diffraction.

For a given source (S), the space is divided into angular regions. They are spherical sectors from the source point defined by the spherical coordinates theta (θ) and phi (ϕ) of a fixed coordinate system located on the source (see Figure 3.26). Using a nomenclature similar to the SVP method, they are called *anxels* as an abbreviation for angular elements. The number of anxels depends

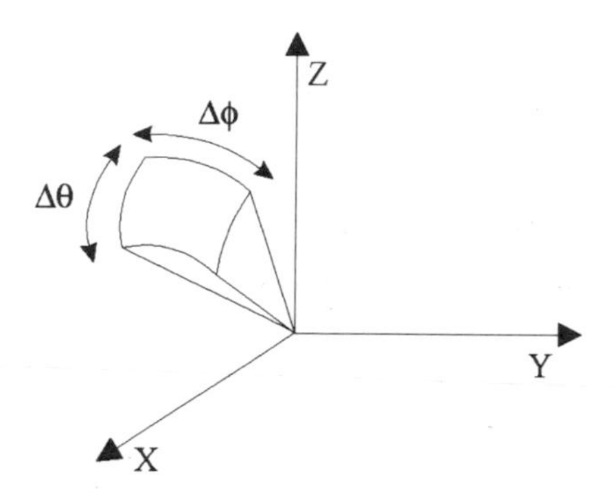

Figure 3.26 Anxel definition.

on the angular margins $\Delta\theta$ and $\Delta\phi$, which define the anxel size. The facets of the scene that lie in each anxel are determined. In other words, the facets are located in the corresponding cell of the θ-ϕ plane (AZB plane) (see Figure 3.27).

Each cell is a representation of an anxel in the AZB plane. Moreover, in each anxel the facets are arranged according to the distance to S. This is calculated as the distance between S and the closer vertex of the facet. All of this information is stored in the so-called AZB matrix. It depends, exclusively, on the source point and on the facets of the environment. As an example, Figure 3.28 shows a simple 2D outdoor scene where a space partitioning in eight anxels has been accomplished. In this case, the anxels degenerate in angular sectors. Table 3.6 shows the storage of the facets in the AZB matrix.

In outdoor scenes, the backface culling test can be applied to remove facets of the AZB matrix. For example, in anxel 1, facets 10 and 11 can be removed. Also, once the facets have been sorted they are checked, starting from

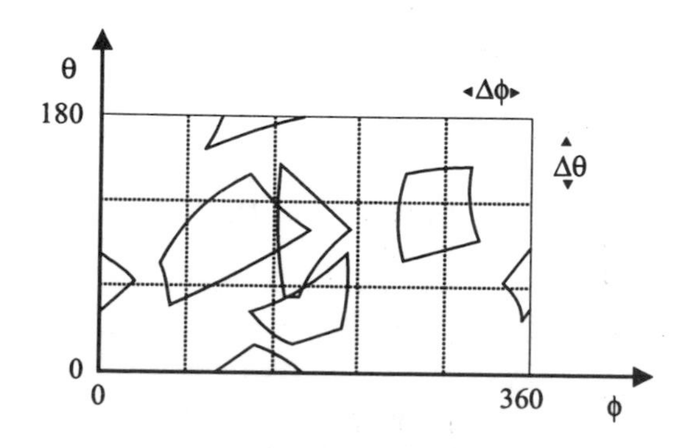

Figure 3.27 Facet storage in the AZB plane.

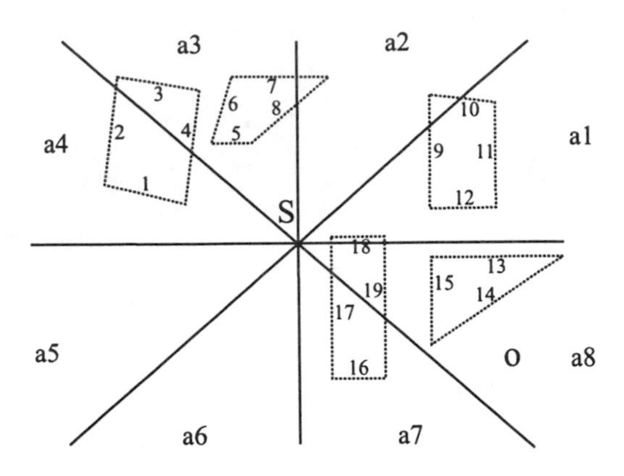

Figure 3.28 Example of an 2D outdoor scene splitting into eight anxels.

the second one, in order to determine if they are shadowed by facets closer to the source. All the facets totally shadowed are removed from the AZB matrix. For example in anxel 8, facet 15 is totally shadowed by facet 17. This procedure of sorting facets in the anxel and discarding the shadowed ones is called the *painter's algorithm* [32], because it resembles the way a painter works: Nearest objects are painted over the farthest ones, shadowing them. The painter's algorithm is used in most *z*-buffer algorithms used in computer graphics.

Table 3.7 shows the simplified AZB matrix information after backface culling and the painter's algorithm have been applied.

For a given observation point (O), the shadowing test determines if the path S–O is hidden by any facet. To complete this task, the spherical coordinates of the observer are calculated and the point is located in the corresponding anxel (a cell in the AZB plane). Only the facets placed in the anxel with distance to S less than the distance S–O can obstruct the path. Moreover, the

Table 3.6
Facet Storage in the AZB of the Scene of Figure 3.28

ANXEL	FACETS
1	17, 18, 19, 9, 12, 11, 10
2	8, 9, 10, 7
3	4, 5, 6, 8, 3, 7, 2
4	1, 4, 2
5	
6	
7	17, 19, 16
8	17, 19, 15, 13, 14

Table 3.7
Facet Storage in the AZB of the Scene of Figure 3.28
(shadowed facets have been removed)

ANXEL	FACETS
1	17, 9, 12
2	8, 9
3	4, 5, 8
4	1, 4
5	
6	
7	17
8	17

facets are tested in an orderly manner, attending to their distance from the source, because the facets closer to S have a higher probability of hiding the path S–O. In the 2D scene of Figure 3.29, only facet 17 is considered in the shadowing test. The reduction in the number of facets tested decreases with the number of anxels, that is, it depends on the values of $\Delta\theta$ and $\Delta\phi$. On the other hand, the number of regions is limited by the available memory size.

For a given number of anxels, the efficiency of the angular partitioning decreases when the size of the scene increases. This is because far away from the source, the area that occupies each anxel will be large so it could contain a large number of facets. But in picocells and microcells, where deterministic 3D ray models are applicable, the size of the scenario is reduced, hence, the

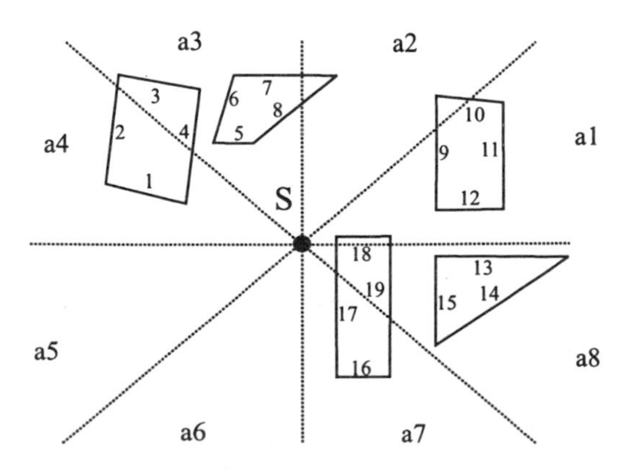

Figure 3.29 The shadowing test. Only facet 17 is tested.

AZB is very efficient. In large cells or macrocells, the AZB can be combined with the BSP or SVP algorithms.

The application of the AZB technique to direct rays is straightforward. We consider the transmitter antenna to be the source (S). For other effects, the application of the AZB is slightly different, as discussed in the following subsections.

Application to Reflected Rays

In this case, the sources are the images of the transmitter antenna (S) with respect to the directly illuminated facets. These facets are determined using the AZB matrix of the direct field computation.

The image points (I) are the sources of the reflected rays. But we must consider that each one of the image points only radiates in the reflection space (RS) (see Figure 3.12). The highest and lowest values of the spherical coordinates (θ, ϕ) of the reflecting facet vertices determine the space where the AZB must be applied. In the 2D representation, this space corresponds with the so-called "AZB rectangle" (see Figure 3.30). As in the direct field analysis, the AZB rectangle is divided into anxels. Then, for each facet seen from I in its AZB rectangle, the anxel or anxels where it lies are determined to fill the AZB matrix. The facets of each anxel are also arranged according to the distance from the source.

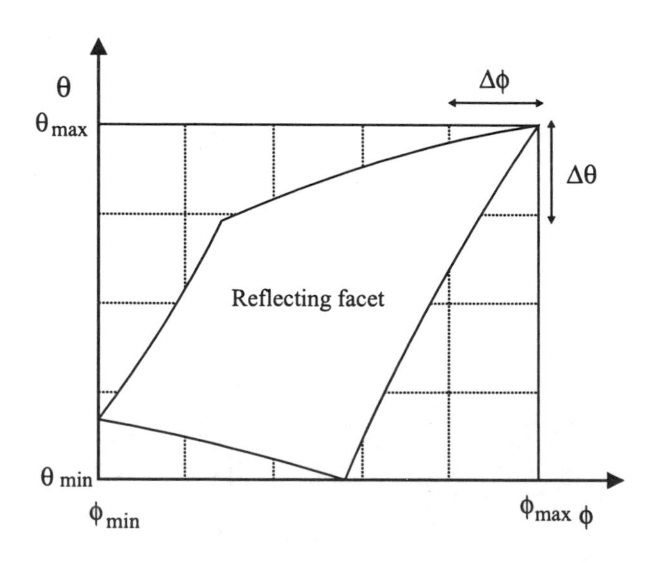

Figure 3.30 The AZB rectangle for an image source. The reflecting facet is seen from the image source as a quadrangle with curved sides (the sides are drawn straight in this figure). The AZB rectangle is split into anxels.

Given an observation point (O), its spherical coordinates (r_i, θ_i, ϕ_i) are calculated. If the point does not lie inside the AZB rectangle there is no reflection in the facet. Otherwise, it is checked to see if it lies in the reflection quadrangle. If it does not, there is no reflection. Otherwise, the anxel of the observation point is found and the facets located on the anxel are tested following the same procedure as in the direct ray. With the above procedure, the possible hiding of the reflected ray is analyzed.

The analysis of the incident ray (from S to the reflection point) is done using the AZB matrix of the direct field, taking the reflection point as the observation point.

Application to Double-Reflected and Higher Order Reflected Rays

Given a first-order image (I), just the facets located in the AZB rectangle can be involved in a double reflection. Then, for each of the above facets, the second-order image (I_2) is determined. For each new source I_2, the previous procedure is performed, that is, the AZB rectangle is obtained as was done for the first-order image (I).

For a given double-reflected ray, the possible hiding is analyzed as follows:

- For the incident ray (path transmitter-reflecting point 1), the AZB matrix of the direct field computation is used considering the first reflection point as the observer.
- For the ray reflecting point 1-reflecting point 2, the AZB matrix of I is used considering the second reflection point to be the observation point.
- For the path reflecting point 2-observer, the AZB matrix of I_2 is used.

If high-order reflections are considered, the number of multiple images can be quite high. On the other hand, in high-order reflections the reflecting space becomes very narrow (the margins of the AZB rectangle become very narrow) and the number of facets to store (and to test) becomes very low. As we will discuss in the comparison between SVP and AZB techniques in most second and higher order reflections cases, the SVP technique will be preferable.

Application to Edge-Diffracted Rays

The application of the AZB algorithm to edge-diffracted rays is slightly different. Now the sources are the points of the edges (infinite points). Moreover, each of them radiates in infinite directions contained in the Keller's cone. The facet arrangement is accomplished in terms of the coordinates β, α of the edge fixed coordinate system (see Figure 3.31) instead of the spherical coordinates

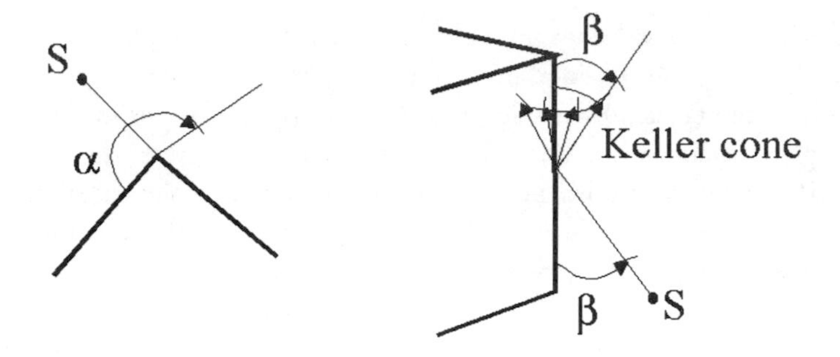

Figure 3.31 Definition of the β and α angular parameters for the AZB of the edge diffraction.

θ, ϕ. Here β is the angle of the Keller's cone in each edge point so it varies along the edge, and α is the angle formed by the diffracted ray and the first facet of the wedge.

Given a source S and an edge, all the diffracted rays can be represented as points in the so-called AZB rectangle of diffraction as shown in Figure 3.32. This rectangle is a 2D representation of the diffraction space. The maximum and minimum values of the edge coordinates (β_{max}, β_{min}, α_{max}, α_{min}) fix the margins of the rectangle. The rectangle is divided into anxels. The facets

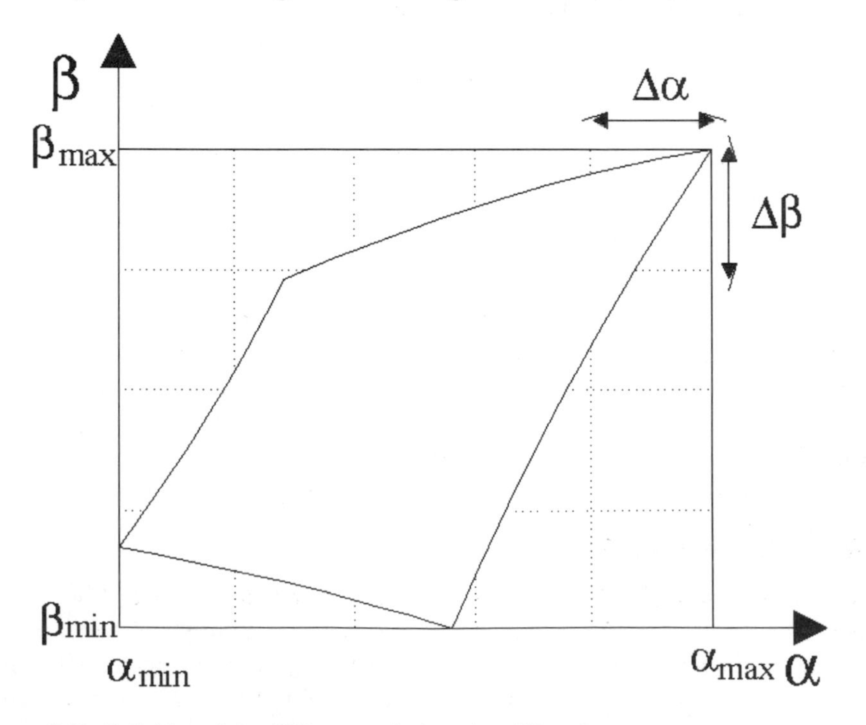

Figure 3.32 Definition of the AZB rectangle for edge diffraction.

of the environment are represented in the AZB rectangle as quadrangles. The vertices of the quadrangles are given by the edge coordinates of the facets vertices.

The information of the AZB rectangles of diffraction depends on the geometry of the environment and on the source location. Therefore, it is independent of the observer point. This information is stored in the so-called AZB matrix of diffraction.

Given an edge and an observation point, its edge coordinates (β_0, α_0) are calculated and the point is located in the AZB rectangle. If it is outside the rectangle margins, there is no diffraction in the edge. Otherwise, the anxel where the point lies is determined. Only the facets stored in the cell are considered in the test of the diffracted ray hiding. The test is made in an orderly way, that is, it begins with the facet closer to the edge as explained for the direct ray case. Obviously, if a facet is farther from the edge than O, it is not tested.

If the diffracted ray is not hidden, the incident ray (source-diffraction point) is analyzed. To complete this task, the AZB matrix of the direct field is used, taking the diffraction point as the observation point.

Application to Reflected-Diffracted Rays

Only the edges located in the reflection space of the facets illuminated by the source are considered. These facets are obtained from the AZB matrix of the direct field. Each one of these facets has its corresponding image of reflection. The AZB matrices of diffraction are calculated as in the single diffraction case, but now the sources are the images of reflection.

With this information, the shadowing test for reflected-diffracted rays is rapidly performed: If the observation point is not in the AZB rectangle of diffraction, there is no reflection-diffraction. Otherwise, the AZB matrix of diffraction is used for the analysis of the diffraction point-observer line. The AZB matrix of reflection is used for the analysis of the reflection point-diffraction point path and the AZB matrix of the direct field is used in the S-reflection point line.

Application to Diffracted-Reflected Rays

Only the edges stored in the AZB matrix of the direct field can be involved in diffraction-reflection, and only the facets stored in the AZB matrices of single diffraction can take part in diffraction-reflection. The diffracted-reflected problem can be reduced to a single diffraction using the following strategy: For each pair edge-facet, the images of the source (S) and edge (image edge) in the reflecting facet are calculated (see Figure 3.33). Now the problem is reduced to a single diffraction on the image edge where the source is the image

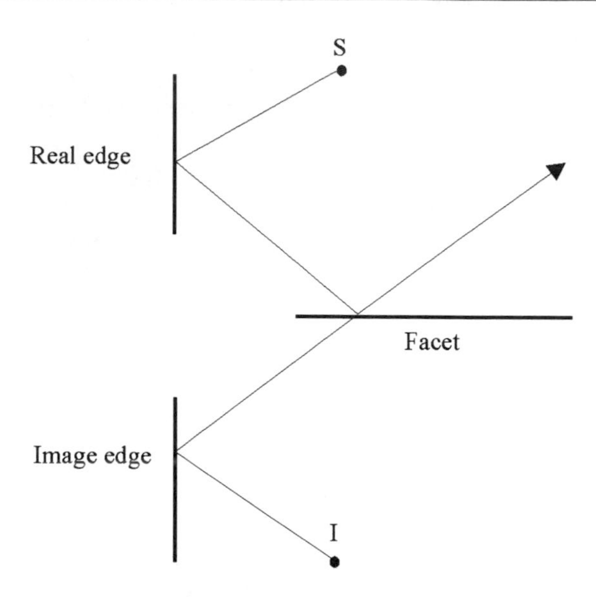

Figure 3.33 The diffraction-reflection is reduced to a simple diffraction calculating the image of the source and the image of the edge with respect to the reflecting facet.

of the transmitter (this was calculated in the single reflected field computation). The AZB matrix of diffraction corresponding to the image edge is calculated and the facets are located in the AZB rectangle following a procedure similar to the one for simple diffraction. The reflecting facet is also located in the AZB of the image edge. With this information, for any observation point, the ray-tracing is rapidly solved. Only the observation points located inside the quadrangle of the image edge and in the same anxels as the reflecting facet are involved in a diffraction-reflection (see Figure 3.34).

The AZB matrix of the image edge is used for the analysis of the path between the reflection point and the observation point. The AZB matrix of the real edge is used for the analysis of the path between the diffraction point and the reflection point. Finally, the line S-diffraction point is analyzed using the AZB matrix of the direct field.

Application to Multiple Interactions Between Edges and Facets

The shadow testing of multiple interactions involving reflections and diffractions can be solved combining the preceding procedures. If the number of diffractions and/or reflections is high, the number of AZB matrices grows and more memory size is necessary. On the other hand, in high-order effects, the margins of the AZB rectangles become very narrow, so the number of facets to be stored (and to be tested) become very low.

The following example can be illustrative of the efficiency of the AZB algorithm; the scene analyzed is depicted in Figure 2.17. It contains about 700

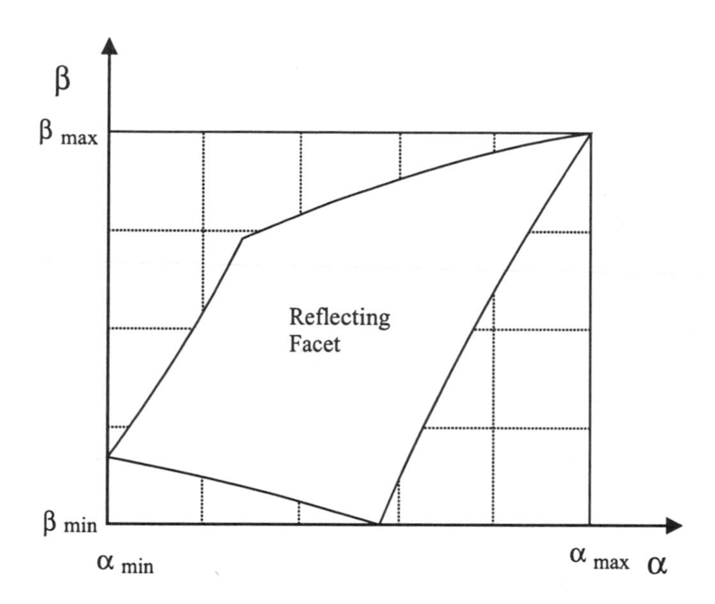

Figure 3.34 AZB rectangle of the image edge. The quadrangle corresponding to the reflecting facet is shown (in the figure the sides are drawn straight).

facets and 1,200 edges. The observation points were located in a mesh of 250 × 250 points, that is, the total number of observation points was 62,500. Simple, double, and some triple effects (those with one effect in the triad being a reflection off the ground) were considered. The computer was a Pentium 120-MHz system with 32 megabytes of RAM memory. The total CPU time was 45 minutes.

A part of the previous scene is represented in Figures 3.7 through 3.9. This 3D scenario contains about 175 facets and 300 edges. Considering the effects indicated in Figure 3.9, the CPU time consumed was about 4 minutes for 10,000 observation points. The computer was the same as in the previous case.

3.5 Rigorous Determination of Diffraction Points

As was shown in previous sections, the determination of reflection points can be efficiently achieved by using the image theory. The determination of the diffraction point in a single diffraction can also be made using the following closed-form expression:

$$\vec{Q}(t) = \vec{p}_1 + t(\vec{p}_2 - \vec{p}_1),\ 0 \le t \le 1 \tag{3.4}$$

with

$$t = \frac{(\vec{p}_2 - \vec{p}_1)}{|\vec{p}_2 - \vec{p}_1|^2} \cdot \left[\frac{(\vec{S}d_o + \vec{O}d_s)}{d_o + d_s} - \vec{p}_1 \right] \tag{3.5}$$

where $\vec{Q}$ is the diffraction point, $\vec{S}$ is the source, $\vec{O}$ is the observation point, $\vec{p}_1$ and $\vec{p}_2$ are the endpoints of the edge, and d_o and d_s are the distance from the observation point and the source to the straight line that contains the edge. The derivation of the preceding expression can be found in Appendix 3C.

When two or more diffractions are involved, the calculation of the diffraction points is not a simple task. This section presents a method for the case of double diffractions but its generalization to the case of multiple diffraction is straightforward.

The diffraction points in a double diffraction can be obtained by applying Fermat's principle, which states that the double-diffraction path is the ray of minimum length that, starting at the source S, and touching both edges in the diffraction points Q_1 and Q_2, finally reaches the observation point O (see Figure 3.35).

The ray length can be written as a function of the edges parameters t_1 and t_2:

$$F(t_1, t_2) = d_1(t_1) + d(t_1, t_2) + d(t_3) \tag{3.6}$$

where d_1, d_2 and d_3 are the partial distances of the ray

$$d_1(t_1) = |\vec{S} - \vec{Q}_1(t_1)| \tag{3.7}$$

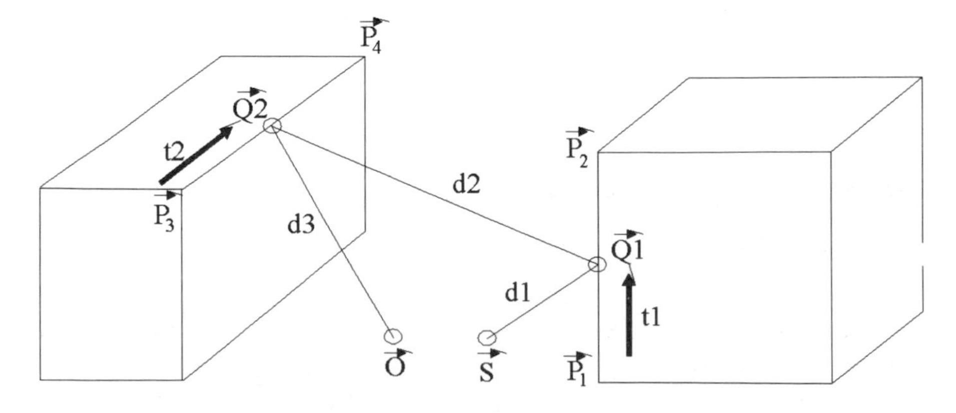

Figure 3.35 Parameters defining a double-diffraction ray path.

$$d_2(t_1, t_2) = |\vec{Q}_1(t_1) - \vec{Q}_2(t_2)| \tag{3.8}$$

$$d_3(t_2) = |\vec{Q}_2(t_2) - \vec{O}| \tag{3.9}$$

$\vec{Q}_1$, $\vec{Q}_2$ are the diffraction points

$$\vec{Q}_1(t_1) = \vec{p}_1 + t_1(\vec{p}_2 - \vec{p}_1) \tag{3.10}$$

$$\vec{Q}_2(t_2) = \vec{p}_3 + t_2(\vec{p}_4 - \vec{p}_3) \tag{3.11}$$

and $\vec{p}_1$, $\vec{p}_2$, $\vec{p}_3$ and $\vec{p}_4$ are the endpoints of the edges.

Minimizing the function F, the edges' parametric coordinates t_1 and t_2 are obtained, and using (3.10) and (3.11), the diffraction points are calculated.

3.6 Rigorous Ray-Facet Intersection Algorithms

The heart of any ray-tracing model is that of the intersection algorithms between rays and the objects of the environment. In a general ray-tracing process the ray object intersection analysis is made many times. In facet models, the rays intersect with facets, so it is necessary to use an efficient ray-facet intersection algorithm [14, 24]. The efficiency of the ray-tracing process depends a lot on the ray-tracing intersection algorithm.

In ray-tracing models for propagation prediction, the rays always have an origin (source) and arrive at an observation point (observer). Therefore, geometrically, rays can be considered as 3D oriented segments, that is, segments with a given direction. The sources can be transmitter antennas, reflection points, or diffraction points. Observation points can be the receiver antennas, reflection points, or diffraction points.

3.6.1 Ray Intersection Algorithms With Arbitrary 3D Facets

The algorithm for arbitrarily shaped facets can be divided into two steps:

1. *Computation of a ray-plane intersection.* First, the plane that contains the facet is determined. Next, the ray-plane intersection point is calculated. If the plane and ray are parallel, there is no intersection point and therefore there is no ray-facet intersection. Otherwise, the ray-plane intersection point is calculated.
2. *Testing to determine whether a point is inside or outside a polygon in the plane.* Knowing the ray-plane intersection point, one can determine

whether the point is inside or outside the polygon. If the point is inside the polygon, there is ray-facet intersection (see Figure 3.36(a)). Otherwise, there is no ray-facet intersection (see Figure 3.36(b)).

Figure 3.37 shows a block diagram of the above procedure.

Determination of the Plane Facet

The equation of a plane

$$Ax + By + Cz + D = 0 \tag{3.12}$$

contains four coefficients that need to be determined. Depending on the available initial information, there are two possible strategies.

If the normal vector to the plane is known (see Appendix 3A), coefficients A, B, and C are known (they are equal to the normal vector components) and the value of D is obtained from any point of the plane (x_1, y_1, z_1):

$$D = -(Ax_1 + By_1 + Cz_1) \tag{3.13}$$

The point of the plane can be any of the facet vertices.

When the normal vector to the plane is unknown, the plane equation can be determined from three noncollinear points on the plane (x_1, y_1, z_1), (x_2, y_2, z_2), (x_3, y_3, z_3):

$$\begin{bmatrix} A \\ B \\ C \end{bmatrix} = \begin{bmatrix} x_1 & y_1 & z_1 \\ x_2 & y_2 & z_2 \\ x_3 & y_3 & z_3 \end{bmatrix} \cdot \begin{bmatrix} -1 \\ -1 \\ -1 \end{bmatrix}, \quad D = 1 \tag{3.14}$$

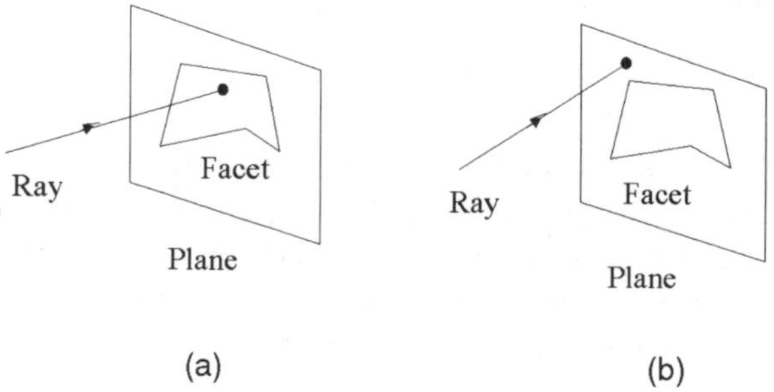

Figure 3.36 Ray-facet intersection.

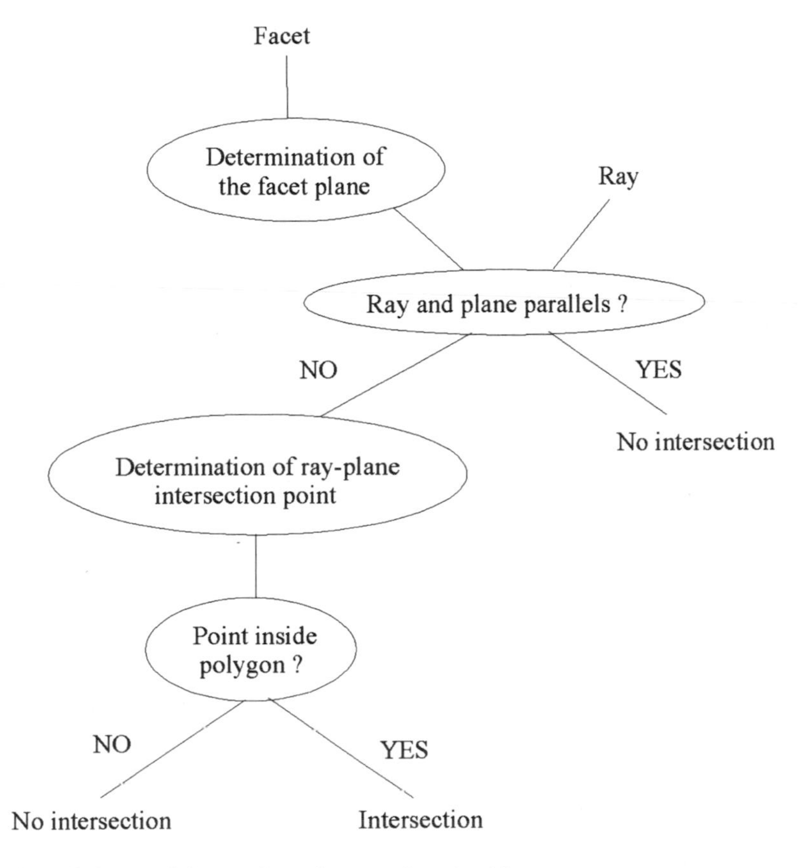

Figure 3.37 Scheme of the ray-facet intersection algorithm.

Ray-Plane Intersection

Let's suppose a ray is defined by its origin $\vec{r}_0$ and its end $\vec{r}_1$. So, the points on the ray:

$$\vec{r}(t) = \vec{r}_0 + t(\vec{r}_1 - \vec{r}_0) \qquad \text{with } 0 \leq t \leq 1 \tag{3.15}$$

Define the plane in terms of its unit normal vector $\hat{n} = (A, B, C)$ and the distance from the coordinate system origin: D.

If $(\vec{r}_1 - \vec{r}_0) \cdot \hat{n} = 0$, the ray is parallel to the plane and no intersection occurs except when the ray is in the same plane and there are infinite intersection points. This last case happens when

$$(\vec{r}_1 - \vec{r}_0) \cdot \hat{n} = 0 \quad \text{and} \quad \vec{r}_0 \cdot \hat{n} + D = 0 \tag{3.16}$$

If $(\vec{r}_1 - \vec{r}_0) \cdot \hat{n} \neq 0$, there is an intersection point between the plane and the line that contains the ray. The parametric coordinate of the ray in the intersection point is

$$t_i = -\frac{\vec{r}_0 \cdot \hat{n} + D}{(\vec{r}_1 - \vec{r}_0) \cdot \hat{n}} \tag{3.17}$$

If $0 \leq t \leq 1$, then the line defined by the ray intersects the plane between the ray's endpoints and hence, ray plane intersection occurs. In this case, the intersection point is given by

$$\vec{r}_i = \vec{r}_0 + t_i \cdot (\vec{r}_1 - \vec{r}_0) \tag{3.18}$$

The block diagram of Figure 3.38 illustrates the procedure just detailed.

Point Inside/Outside a Polygon

A number of different methods are available to solve this problem. The method presented here [15, 25] is based on the so-called "Jordan curve" theorem. It is valid for arbitrary-shaped polygons.

The givens are a polygon and a point $\vec{r}_i$ located on a plane. A ray is shot from $\vec{r}_i$ in an arbitrary direction on the plane. Counting the number of polygon sides crossed, one can determine whether $\vec{r}_i$ is inside or outside the polygon by using the following rule: If the number of crossings is odd, the point is inside the polygon, otherwise; it is outside. Figure 3.39 shows an example.

Given a testing point $\vec{r}_i$, the possible intersection ray segment in the plane is checking N times, N being the number of vertices of the polygon.

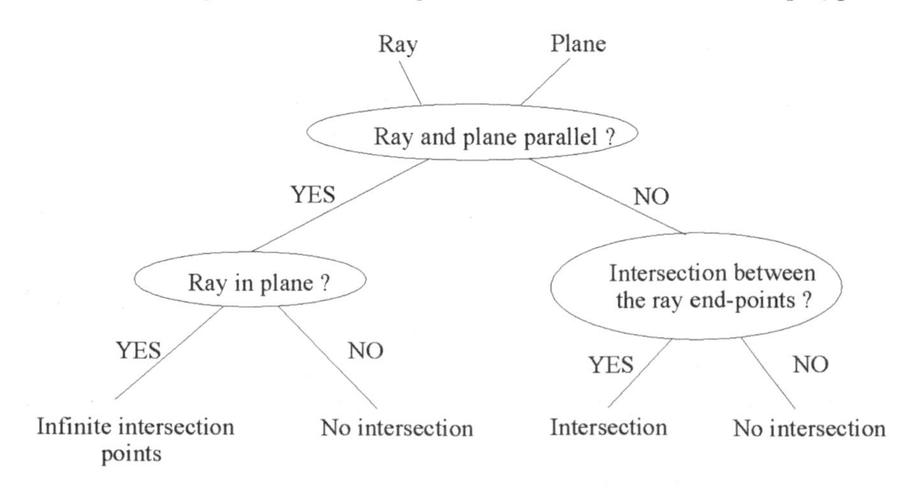

Figure 3.38 Scheme of the ray plane intersection algorithm.

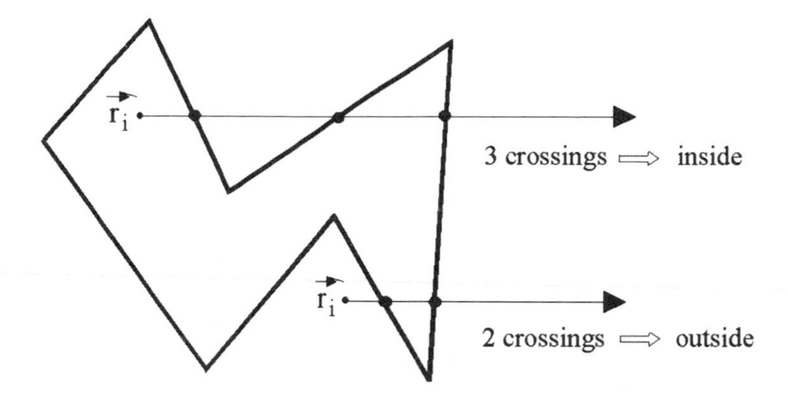

Figure 3.39 Point inside or outside an arbitrary facet.

Ray-Segment Intersection in a Plane

Let's consider a 2D coordinate system on the plane. As stated earlier, the ray can be arbitrarily orientated. Therefore, in order to simplify (and optimize) the algorithm, a ray is chosen parallel to any of the coordinate axes. Let's suppose a ray is parallel to the *X* axis, with its origin in $\vec{r}_i = (x_i, y_i)$. Given a segment with endpoints $\vec{r}_{s1} = (x_{s1}, y_{s1})$ and $\vec{r}_{s2} = (x_{s2}, y_{s2})$, its parametric equation is

$$\vec{r} = \vec{r}_{s1} + t(\vec{r}_{s2} - \vec{r}_{s1}),\ 0 \leq t \leq 1 \tag{3.19}$$

A ray segment intersection (see Figure 3.40) occurs if the following conditions are fulfilled:

$$y_{s1} \neq y_{s0} \tag{3.20}$$

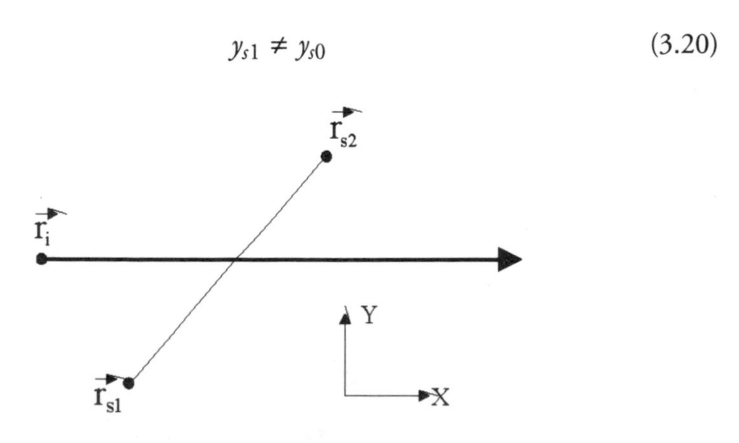

Figure 3.40 Ray-segment intersection in a plane.

$$0 \le t_s \le 1, \qquad \text{where } t_s = \frac{y_i - y_{s1}}{y_{s2} - y_{s1}} \tag{3.21}$$

$$x_i \le x_{s1} + t_s(x_{s2} - x_{s1}) \tag{3.22}$$

The first condition ensures that the ray and the segment are not parallel, the second ensures that the segment is cut between the endpoints, and the third condition ensures that the ray is not cut behind its origin.

A particular case occurs when the segment is contained (totally or partially) by the ray. Therefore, there are infinite intersection points. This happens when the following conditions are fulfilled:

$$y_{s1} = y_{s2} = y_i \tag{3.23}$$

$$x_i < x_{s1} \text{ or } x_i < x_{s2} \tag{3.24}$$

The first condition ensures that the segment lies on the line that contains the ray. The second condition ensures that there is a portion of the segment ahead of the origin of the ray.

3.6.2 Ray Intersection Algorithms With Particular 3D Facets

In urban and indoor scenes, many of the facets are vertically oriented: facets of walls, facets representing doors, the majority of the facets from windows, etc. This geometric property can be used to simplify the ray-facet intersection algorithm for this kind of facet. To do this, a coordinate system with the Z axis parallel to these facets is chosen. The plane X-Y (horizontal plane) will be perpendicular to such facets.

In the horizontal plane the wall facets are represented as arbitrary oriented segments. Therefore, the ray-facet intersection analysis can be divided into two steps. First, the ray is projected onto the plane. The intersection projected ray-facet is achieved in the plane. So, this first step is reduced to a ray-segment intersection in a 2D scene.

Second, if the intersection point has been found, the Z component of the intersection point is calculated. Finally, by comparing the previous value with the Z components of the facet vertices, it is easy to determine if the intersection point lies on the facet.

Intersection of Segments in a Plane

As stated earlier, the rays can be assumed to be "oriented segments," so the problem is to analyze the intersection of two segments in the plane.

Given two segments with endpoints $\vec{r}_{s1} = (x_{s1}, y_{s1})$, $\vec{r}_{s2} = (x_{s2}, y_{s2})$ and $\vec{r}_{s3} = (x_{s3}, y_{s3})$, $\vec{r}_{s4} = (x_{s4}, y_{s4})$, respectively.

Let the coefficients be

$$A = y_{s2} - y_{s1}, \quad A' = y_{s4} - y_{s3}, \quad B = x_{s1} - x_{s2}, \quad B' = x_{s3} - x_{s4},$$
$$C = y_{s1}x_{s2} - y_{s2}x_{s1}, \quad C' = y_{s3}x_{s4} - y_{s4}x_{s3} \tag{3.25}$$

If $AB' - A'B = 0$, the segments are parallel and there is no intersection. Otherwise, the lines that contain the segments have an intersection point (x_i, y_i) given by:

$$x_i = \frac{BC' - B'C}{AB' - A'B}, \qquad y_i = \frac{CA' - C'A}{AB' - A'B} \tag{3.26}$$

Finally, such a point belongs to both segments if the following conditions are verified:

$$0 \le \frac{x_i - x_{s1}}{x_{s2} - x_{s1}} \le 1 \quad \text{or} \quad 0 \le \frac{y_i - y_{s1}}{y_{s2} - y_{s1}} \le 1 \tag{3.27}$$

$$0 \le \frac{x_i - x_{s3}}{x_{s4} - x_{s3}} \le 1 \quad \text{or} \quad 0 \le \frac{y_i - y_{s3}}{y_{s4} - y_{s3}} \le 1 \tag{3.28}$$

References

[1] Sou, C. K., O. Landron, and M. J. Feuerstein, "Characterization of Electromagnetic Properties of Building Materials for Use in Site-Specific Propagation Prediction," Mobile Portable Radio Research Group Technical Report No. 92-12, Virginia Tech, 1992.

[2] Landron, O., M. J. Fuerstein, and T. S. Rappaport, "A Comparison of Theoretical and Empirical Reflection Coefficients for Typical Exterior Wall Surfaces in a Mobile Radio Environment," *IEEE Trans. on Antennas and Propagation,* Vol. 44, No. 3, Mar. 1996, pp. 341–351.

[3] Honcharenko, W., and H. L. Bertoni, "Transmission and Reflection Characteristics at Concrete Block Walls in the UHF Band Proposed for Future PCS," *IEEE Trans. on Antennas and Propagation,* Vol. 43, No. 2, Feb. 1994, pp. 232–239.

[4] Lawton, M. C., and J. P. McGeehan, "The Application of a Deterministic Ray Launching Algorithm for the Prediction of Radio Channel Characteristics in Small-Cell Environments," *IEEE Trans. on Vehicular Technology,* Vol. 43, No. 4, Nov. 1994, pp. 955–969.

[5] Von Hipple, A. R., *Dielectric Materials and Applications,* New York: The MIT Press and Wiley, 1954.

[6] *American Institute of Physics Handbook,* 3rd ed., New York: McGraw-Hill, 1972.

[7] Balanis, C. A., *Advanced Engineering Electromagnetics,* New York: Wiley and Sons, 1989.

[8] Maguire, D. J., M. F. Goodchild, and D. W. Rhind, *Geographical Information Systems,* Essex, UK: Longman Scientific & Technical, 1991.

[9] Kürner, T., D. J. Cichon, and W. Wiesbeck, "Concepts and Results for 3D Digital Terrain-Based Wave Propagation Models: An Overview," *IEEE J. on Selected Areas in Communications,* Vol. 11, No. 7, Sep. 1993, pp. 1002–1012.

[10] Tan, S. Y., and H. S. Tan, "A Microcellular Communications Propagation Model Based on the Uniform Theory of Diffraction and Multiple Image Theory," *IEEE Trans. on Antennas and Propagation,* Vol. 44, No. 10, Oct. 1996, pp. 1317–1326.

[11] Schaubach, K. R., and N. J. Davis, "Microcellular Radio-Channel Propagation Prediction," *IEEE Antennas and Propagation Mag.,* Vol. 36, No. 4, Aug. 1994, pp. 25–34.

[12] Rappaport, T. S., S. Y. Seidel, and K. R. Schaubach, "Site-Specific Propagation for PCS System Design," in *Wireless Personal Communications,* M. J. Feuerstein, T. S. Rappaport, eds., Boston: Kluwer Academic Publishers, 1993, pp. 281–315.

[13] IGES/PDES Organization, "The Initial Graphics Exchange Specification (IGES) Version 5.1, Sep. 1991.

[14] *AutoCAD Release 10 Reference Manual,* 1988.

[15] Glassner, A. S. (Ed.), *An Introduction to Ray Tracing,* San Diego, CA: Academic Press, 1989.

[16] Foley, J. D., A. van Dam, S. K. Feiner, and J. F. Hughes, *Computer Graphics. Principles and Practice,* 2nd ed., New York: Addison-Wesley, 1995.

[17] Cátedra, M. F., J. Pérez, A. González, O. Gutiérrez, and F. Saez de Adana, "Fast Computer Tool for the Analysis of Propagation in Urban Cells," *Proc. of Wireless Communications Conference,* Boulder, CO, Aug. 11-13, 1997, pp. 240–245.

[18] Cátedra, M. F., J. Pérez, F. Saez de Adana, and O. Gutiérrez, "Efficient Ray-Tracing Techniques for 3D Analysis of Propagation in Mobile Communications. Application to Picocell and Microcell Scenarios," *IEEE Antennas and Propagation Mag.*, Vol. 40, No. 2, April 1998, pp. 15–28.

[19] Seidel, S. Y., and T. S. Rappaport, "Site-Specific Propagation Prediction for Wireless in Building Personal Communication System Design," *IEEE Trans. on Vehicular Technology,* Vol. 43, No. 4, Nov. 1994, pp. 879–891.

[20] McKown, J. W., and R. L. Hamilton, "Ray Tracing as a Design Tool for Radio Networks," *IEEE Network Mag.,* Nov. 1991, pp. 27–30.

[21] Lee, S. W., J. Baldauf, and L. Lin, "CAD-Based RCS Computation," *Proc. 1989 IEEE Antennas and Propagation Society Symposium,* June 1989.

[22] Ling, H., R. Chou, and S. W. Lee, "Shooting and Bouncing Rays: Calculating the RCS of an Arbitrary Shaped Cavity," *IEEE Trans. on Antennas and Propagation,* Vol. 37, Feb. 1989, pp. 194–205.

[23] Lee, S. W., H. Ling, and R. Chou, "Ray-Tube Integration in Shooting and Bouncing Ray Method," *Microwave Opt. Technol. Lett.,* Vol. 1, No. 8, Aug. 1988, pp. 286–289.

[24] Honcharenko, W., H. L. Bertoni, J. L. Dailing, J. Quian, and H. D. Yee, "Mechanism Governing UHF Propagation on Single Floors in Modern Office Buildings," *IEEE Trans. on Vehicular Technology,* Vol. 41, No. 4, Nov. 1992.

[25] Rogers, D. F., *Procedural Elements for Computer Graphics,* New York: McGraw-Hill, 1985.

[26] Arvo, J., and D. Kirk, "A Survey of Ray Tracing Acceleration Techniques" in *An Introduction to Ray Tracing,* A. S. Glassner, ed., San Diego, CA: Academic Press, 1989, pp. 201–262.

[27] Fuchs, H., "On Visibility Surface Generation by a priori Tree Structures," *Computer Graphics,* Vol. 14, No. 3, July 1980, pp. 124–133.

[28] Fujimoto, A., C. G. Perrott, and K. Iwata, "ARTS: Accelerated Ray-Tracing System," *IEEE Computer Graphics and Applications,* Vol. 6, No. 4, Apr. 1986, pp. 16–26.

[29] Rubin, H. S., and T. Whitted, "Three-Dimensional Representation for Fast Rendering of Complex Scenes," *Computer Graphics, Proc. SIGGRAPH 80,* 1980, pp. 110–116.

[30] Kaplan, M., "Space-Tracing: A Constant Time Ray-Tracer," Course notes from tutorial "State of the Art in Image Synthesis," *SIGGRAPH 85,* 1985.

[31] Hines, E. A., and D. P. Greenberg, "The Light Buffer: A Shadow-Testing Accelerator," *IEEE CG&A,* Sep. 1986, pp. 6–16.

[32] Newell, M. E., R. G. Newell, and T. L. Sancha, "A Solution to the Hidden Surface Problem," *Proc. of ACM National Conference,* 1972, pp. 443–450.

Appendix 3A: Normal Vector of an Arbitrary Facet

The normal vector $\hat{n}$ of an arbitrary facet can be obtained using the following expression:

$$\hat{n} = \frac{1}{2A}[\vec{a}_1 \times \vec{a}_2 + \vec{a}_2 \times \vec{a}_3 + \ldots + \vec{a}_{N-1} \times \vec{a}_N + \vec{a}_N \times \vec{a}_1] \tag{3A.1}$$

where N is the number of vertices, $\{\vec{a}_i\}$ are the vertices, and A is the facet area. Usually, the facet area is unknown, so in order to obtain $\hat{n}$, the expression inside the bracket is calculated and the resulting vector is normalized.

In (3A.1) the direction of the normal vector follows the right-hand rule.

Appendix 3B: Backface Culling

In objects with a closed surface, when there is a ray-object intersection, the ray clips the surface at least twice. When the normal vectors of the surface are outward from the object, the intersection point closer to the ray origin always fulfill

$$(\vec{r}_1 - \vec{r}_0) \cdot \hat{n} < 0 \tag{3B.1}$$

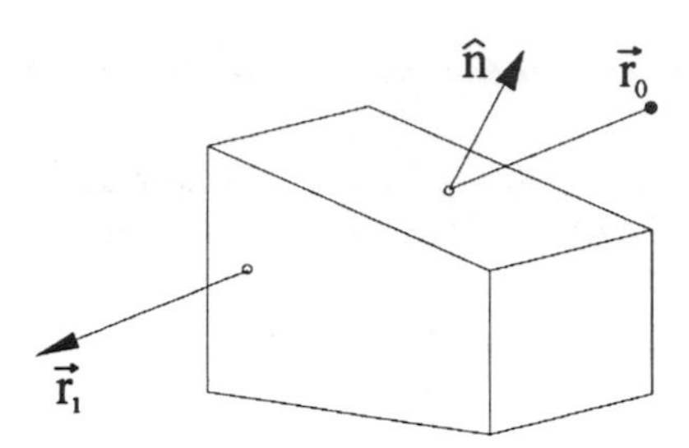

Figure 3B.1 Application to the backface culling.

where $\vec{r}_0$ is the ray origin, $\vec{r}_1$ is the ray end, and $\hat{n}$ is the normal vector of the surface in the intersection point (see Figure 3B.1).

Therefore, if the object surface is modeled with plane facets, only the facets which fulfill (3B.1) must be taken into account in the shadowing test. As a consequence, an important reduction in CPU time can be attained.

Back-face culling criterion can be used in the search for reflection and diffraction points as well because they always lie on facets which satisfy equation (3B.1).

Appendix 3C: Calculation of the Diffraction Point in an Arbitrarily Orientated Straight Edge

Figure 3C.1 shows the ray tracing of a diffraction in a straight edge, where $\vec{Q}$ is the diffraction point, $\vec{S}$ is the source, $\vec{O}$ is the observation point and $\vec{p}_1$ and $\vec{p}_2$ are the end-points of the edge. The edge points can be expressed in a parametric form as follows:

$$\vec{Q}(t) = \vec{p}_1 + t(\vec{p}_2 - \vec{p}_1), \qquad \text{for } 0 \leq t \leq 1 \tag{3C.1}$$

In Figure 3C.1, d_o and d_s represent the distances from the observation point and the source to the straight line which contains the edge where r_s and r_o are the distances along the straight line between the source and the diffraction point and between the observation point and the diffraction point, respectively.

These distances can be written as follows:

$$d_s = \sqrt{(\vec{S} - \vec{p}_1) \cdot (\vec{S} - \vec{p}_1) - \left[\frac{(\vec{p}_2 - \vec{p}_1) \cdot (\vec{S} - \vec{p}_1)}{|\vec{p}_2 - \vec{p}_1|}\right]^2} \tag{3C.2}$$

$$d_o = \sqrt{(\vec{O} - \vec{p}_1) \cdot (\vec{O} - \vec{p}_1) - \left[\frac{(\vec{p}_2 - \vec{p}_1) \cdot (\vec{O} - \vec{p}_1)}{|\vec{p}_2 - \vec{p}_1|}\right]^2} \tag{3C.3}$$

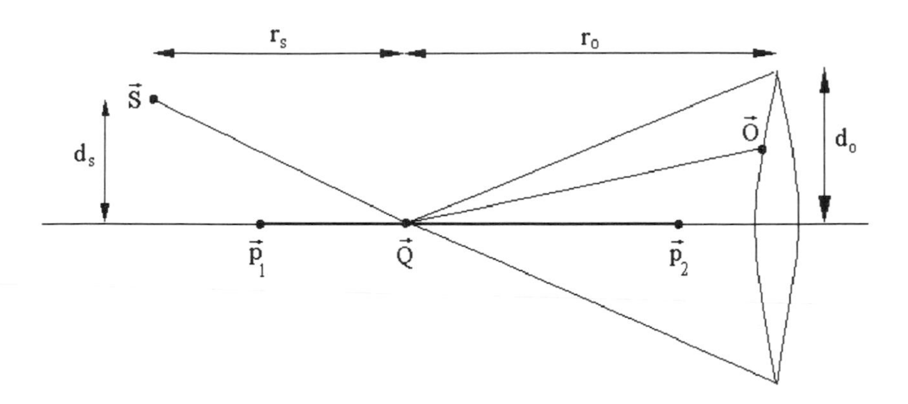

Figure 3C.1 Diffraction at a straight edge.

$$r_s(t) = \frac{(\vec{Q}(t) - \vec{S}) \cdot (\vec{p}_2 - \vec{p}_1)}{|\vec{p}_2 - \vec{p}_1|} \tag{3C.4}$$

$$r_o(t) = \frac{(\vec{O} - \vec{Q}(t)) \cdot (\vec{p}_2 - \vec{p}_1)}{|\vec{p}_2 - \vec{p}_1|} \tag{3C.5}$$

The points of the Keller's cone verify

$$\frac{r_s}{d_s} = \frac{r_o}{d_o} \tag{3C.6}$$

Introducing the expressions (3C.4–5) in (3C.6), and considering equation (3C.1), one can extract out the parameter t:

$$t = \frac{(\vec{p}_2 - \vec{p}_1)}{|\vec{p}_2 - \vec{p}_1|^2} \cdot \left[\frac{(\vec{S}d_o + \vec{O}d_s)}{d_o + d_s} - \vec{p}_1 \right] \tag{3C.7}$$

If the value of the parameter t is between 0 and 1, there is a diffraction point. Then, introducing the value of t in (3C.1), one obtains said diffraction point.

4

Empirical and Semiempirical Path Loss Models

If the received power (or field strength) at the receiver is represented as a function of the transmitter-mobile distance, one can see rapid fluctuations when the distance varies a fraction of a wavelength. These variations of the signal level can be as much as 40 dB. This phenomenon is called fast fading, or small-scale fading.

Fast fading is a consequence of multipath propagation, which occurs when the signals reach the receiver from different paths. The paths followed by the individual waves have different lengths. The total received signal is the coherent sum of the individual signals arriving with different phases and, therefore, producing the fast fading effect. Fast fading can also be observed in a fixed receiver location when the signal level is represented as a function of the frequency.

Another consequence of multipath propagation is the time dispersion phenomenon. Due to the different path lengths, the multipath components do not reach the receiver at the same time, producing time dispersion of the received signals.

The effects of multipath propagation are predicted by the so-called small-scale models or multipath models. Chapter 2 deals with one of these methods.

Another parameter used to characterize the mobile propagation channel is the local average receiver signal, also known as the sector average. This is obtained by averaging the received signal over tracks of several wavelengths or small areas with a radius of various wavelengths. The length of the tracks can vary between 5 (in picocells) and 40 wavelengths (in macrocells) which, at frequencies of the order of 1 GHz, corresponds to distances between approximately 1 and 10m [1].

When the receiver moves along a trajectory, the local average signal suffers fluctuations known as slow fading or large-scale signal variations. These variations of the local average are a consequence of the attenuation of the signal with the distance between the transmitter and receiver and the presence of the obstacles in the environment. The rate of the slow fading variations is of the order of the size of the obstacles surrounding the receiver. Therefore, in rural environments, the local average shows a smoother variation than in urban scenarios.

The so-called large-scale propagation models predict the local average signal level considering the influence of environmental obstacles. This chapter deals with this type of propagation model with special attention paid to the urban and indoor models at the range of frequencies used by mobile communications. The goal of the chapter is not to present an exhaustive revision of the methods. Some of the most commonly employed models are presented as examples.

4.1 Path Loss

Path loss or propagation loss (L) is the parameter commonly used to characterize the local average signal in mobile channels. It is defined as the relationship between the transmitted power (P_t) of the transmitter (Tx) antenna and the received power (P_r) by the receiver (Rx) antenna. In most cases, it is expressed in dB, so

$$L = 10 \log \frac{P_t}{P_r} = P_t(\text{dB}) - P_r(\text{dB}) \tag{4.1}$$

In the definition of path loss, the antenna gains may or may not be included. For example, in free-space propagation, the received power at a distance d from the Tx is given by the Friis equation:

$$P_r(d) = \frac{P_t G_t G_r \lambda^2}{(4\pi)^2 d^2} \tag{4.2}$$

where λ is the wavelength and G_t and G_r are the Tx and Rx antenna gains, respectively. Therefore, the path loss for free-space propagation is given by

$$L = 10\log\left(\frac{(4\pi)^2 d^2}{G_t G_r \lambda^2}\right) = K + 20\log d + 20\log f - 10\log G_t - 10\log G_r \tag{4.3}$$

where f is the frequency and K is a constant that depends on the units of d and f. For example, when d is in kilometers and f is in megahertz, $K = 32.44$. When the antenna gains are not included in the path loss definition, (4.3) reduces to

$$L = 10\log\left(\frac{(4\pi)^2 d^2}{G_t G_r \lambda^2}\right) = K + 20\log d + 20\log f \tag{4.4}$$

The Friis equation is valid in the far-field region of the Tx antenna, that is, when

$$d \gg D \wedge d \gg \lambda \wedge d > \frac{2D^2}{\lambda} \tag{4.5}$$

which is the situation in mobile communications.

Path loss can also be written in terms of the electric field strength $|E|$ instead of the received power. This is related by the following expression:

$$|E| = \sqrt{\frac{P_r \eta 4\pi}{G_r \lambda^2}} \tag{4.6}$$

where η is the characteristic wave impedance of the free space: $120\pi\Omega$.

4.2 Classification of the Propagation Models

Attending to their nature, the propagation models can be classified [2] as

- Empirical models;
- Semiempirical or semideterministic models;
- Deterministic models.

The empirical models are described by equations derived from statistical analysis of a large number of measurements. These methods are simple and do not require detailed information about the environment. They are also easy and fast to apply because the estimation is usually obtained from closed expressions. On the other hand, they cannot provide a very accurate estimation of the path loss.

The deterministic models are based on the application of well-known electromagnetic techniques to a site-specific description of the environments. The environmental description is obtained from building and terrain databases. From these data, one can obtain a description of the scenario in terms of canonical or primitive entities which can be managed efficiently by the electromagnetic theory used. Different degrees of accuracy can be found in the environmental description. Most of the deterministic models are based on ray-tracing electromagnetic methods. These methods are presented in Chapters 2 and 3.

Finally, semiempirical or semideterministic models are based on the equations derived from the application of deterministic methods to generic urban or indoor models. Sometimes, the equations have been corrected experimentally in order to improve their agreement with the measurements. The resulting equations are functions of the characteristics of the areas surrounding the antennas and certain specific characteristics of the scenario. These methods require more detailed information about the environment than the empirical methods but not as much as the deterministic models. With the required data known, they are easy and fast to apply because, as with the empirical models, the results are obtained from closed expressions.

This chapter focuses on empirical and semiempirical propagation models, while Chapters 2 and 3 deal with the deterministic models.

Due to the diversity of the environments where mobile communications occur, each propagation model is devised for a specific type of environment. Therefore, the propagation models can be classified according to the scenario to which they are applied. Three generic categories of environments (cells) can be considered: (1) macrocells, (2) microcells, and (3) picocells.

The macrocells are large areas with the transmitter antenna well above the surrounding buildings. The macrocells occupy a radius between 1 and 30 km from the transmitter. Usually, there is no direct visibility between the transmitter and the receiver.

The microcells occupy smaller areas with a radius between 0.1 and 1 km from the transmitter. The coverage area is not expected to be circular. The transmitter antenna can be above, below, or at the same level as the surrounding buildings. Usually, two situations are distinguished as a function of the relative locations of the transmitter and receiver antenna and the environmental obstacles: LOS (line-of-sight) situations and NLOS (non-line-of-sight) situations.

Finally, the typical dimensions of the picocells are between 0.01 and 0.1 km. Two types of picocells can be considered: indoor and outdoor. The transmitter antenna is below the rooftop level or in a building. In both indoor and outdoor cases, the LOS and NLOS are usually considered separately.

In general, there is a relationship between these three types of models and the types of environments for which they are suitable. Due to their inherent empirical nature, the empirical and semiempirical models are suitable for macrocells with homogeneous characteristics. The semiempirical models are also suitable for homogeneous microcells where the parameters considered by the method characterize the whole scenario well. The deterministic models are suitable for microcells and picocells independently of their shapes. But they are not adequate for macrocells because the CPU times required in such environments make these techniques inefficient.

4.3 Macrocell Propagation Models

4.3.1 Allsebrook Model

The Allsebrok model is an empirical method [3, 4] developed from measurements in British cities between 75 and 450 MHz. The median path loss is given by

$$L = L_f + \sqrt{L_d^2 + (L_p - L_f)^2} + L_b + \gamma \qquad (4.7)$$

where L_f is the free-space path loss, L_p is the so-called plane-earth path loss due to the wave reflection on the ground (assumed flat), L_d accounts for the terrain losses, L_b is the diffraction loss due to buildings, and γ is a correction factor to extrapolate the measurements to the UHF band. The free-space path loss is given by (4.4) and the plane-earth path loss is given by

$$L_p = 120 - 20\log h_t - 20\log h_r + 40\log d \qquad (4.8)$$

where h_t and h_r are the height of the Tx and Rx antennas, respectively. The term L_d is calculated using a multiple knife-edge diffraction model recommended in [5]. For the diffraction losses associated with the buildings, Deslile et al. [6] proposed the following expression:

$$L_b = 10\log\left[\frac{(h_0 - h_r)\sqrt{2\sin\phi}}{548\sqrt{wf}}\right] \qquad (4.9)$$

where h_0 is the average height of the buildings in the proximity of the Rx, w is the width of the street where the receiver is located, and ϕ is the angle formed by the Tx-Rx line and the axis of the street.

4.3.2 Okumura-Hata Model

The Okumura-Hata model is an empirical model developed from measured data obtained in Tokyo at frequencies of 150, 450, and 900 MHz [7]. The model provides closed expressions for the median path loss for different environments [8]. For an urban environment, the median path loss is

$$L_u = 69.55 + 26.16 \log f - 13.82 \log h_t - a(h_m) + (44.9 - 6.55 \log h_t) \log d \tag{4.10}$$

where

f: frequency in the range 150 MHz $\leq f \leq$ 1500 MHz;
h_t: effective Tx antenna height in the range 30m $\leq h_t \leq$ 200m;
h_m: Rx antenna height in the range 1m $\leq h_m \leq$ 10m;
d: distance between Tx and Rx in the range 1 km $\leq d \leq$ 20 km.

The term $a(h_m)$ is a correction factor that depends on the effective Rx antenna height, the frequency, and the size of the coverage area. For small and medium-sized cities,

$$a(h_m) = (1.1 \log f - 0.7) h_m - (1.56 \log f - 0.8) \tag{4.11}$$

For large cities,

$$a(h_m) = 8.29(\log 1.54 h_m)^2 - 1.1 \quad \text{when } f \leq 300 \text{ MHz} \tag{4.12}$$

$$a(h_m) = 3.2(\log 11.75 h_m)^2 - 4.97 \quad \text{when } f \geq 300 \text{ MHz} \tag{4.13}$$

In a suburban environment,

$$L_{su} = L_u - 2\left[\log\left(\frac{f}{28}\right)\right]^2 - 5.4 \tag{4.14}$$

In a rural environment,

$$L_r = L_u - 4.78 \log^2 f + 18.33 \log f - 4.94 \tag{4.15}$$

This model works adequately for large cell mobile systems, but not for cells with a radius of the order of 1 km, which can be found in personal communications systems. Additional improvements on the model can be found in [9].

4.3.3 COST 231-Hata Model

This is an empirical model obtained from the Okumura-Hata model to cover the frequency range from 1500 to 2000 MHz [10]. The range for the rest of the parameters is the same as in the Hata model. The path loss is given by

$$\begin{aligned} L_u = &\ 46.3 + 33.9 \log f - 1.82 \log h_t - a(h_r) \\ &+ (44.9 - 6.55 \log h_t) \log d + C_M \end{aligned} \tag{4.16}$$

where $C_M = 3$ dB in metropolitan centers and 0 dB otherwise.

4.3.4 Lee Model

This is an empirical model for urban and suburban macrocells that predicts the power received P_r in dBm as follows [1]:

$$P_r = A - B \log d - n \log\left(\frac{f}{900}\right) + 10 \log \alpha \tag{4.17}$$

where d is the distance in kilometers between the Tx and the Rx antennas and f is the frequency in MHz. The parameters A and B depend on the environmental characteristics. These have been determined from measurements in several cities with different characteristics that can be used as models.

$$A, B = \begin{cases} 53.9, 38.4; & \text{Suburban area} \\ 62.5, 36.8; & \text{Philadelphia} \\ 55.2, 43.1; & \text{Newark} \\ 77.8, 30.5; & \text{Tokyo} \end{cases}$$

The factor n has the following values

$$n = \begin{cases} 2, & \text{in a suburban area and } f < 450 \text{ MHz} \\ 3, & \text{in an urban area and } f > 450 \text{ MHz} \end{cases} \tag{4.18}$$

The parameter α is given by

$$\alpha = \frac{h_t^2 h_r^m p_t g_t g_r}{3660} \tag{4.19}$$

where h_t and h_r are the Tx and Rx antenna heights (in meters), respectively, p_t is the transmitted power in watts, g_t and g_r are the gain of the Tx and Rx antennas, respectively, and m depends on the Rx antenna height in the following way:

$$m = \begin{cases} 2, & \text{when } h_r > 10\text{m} \\ 1, & \text{when } h_r < 3\text{m} \end{cases} \tag{4.20}$$

4.3.5 Ibrahim and Parsons Models

These are two empirical models derived from measurements in London [11, 12] at three different frequencies (168, 445, and 900 MHz). The transmitter antenna was 46m above local ground. The data were measured in different square areas of 500m per side. For each area, two coefficients related to the urbanization were determined: the land usage factor (F) and the degree of urbanization factor (U). The L term is the percentage of the area covered by buildings, whereas U is the percentage of the area occupied by buildings with four or more floors. Because it is difficult to obtain information about parameter U, in practice, it is taken into account in highly urbanized areas exclusively.

The first model was derived from multiple regression analysis of the measured data. The resulting equation for the median path loss is

$$L = -20\log(0.7h_b) - 8\log h_m + \frac{f}{40} + 26\log\frac{k}{40} - 86\log\left(\frac{f+100}{156}\right) + \left[40 + 14.15\log\left(\frac{f+100}{156}\right)\right]\log d + 0.265L - 0.37H + K \tag{4.21}$$

where $K = 0.087U - 5.5$ for highly urbanized areas; otherwise $K = 0$. The h_b term represents the height of the base station antenna, h_m is the height of the mobile antenna, f is the frequency, d is the distance transmitter-receiver, and H is the difference in average ground height between the area containing the base station and the area containing the receiver.

The second model expresses the median path loss as a correction, obtained from the measurements, of the fourth-power range dependence law. Therefore, the median path loss is given by

$$L = 40\log d - 20\log(h_b h_m) + \beta \tag{4.22}$$

The correction term β is given by

$$\beta = 20 + \frac{f}{40} + 0.18L - 0.34H + K \tag{4.23}$$

where $K = 0.094U - 5.9$ for highly urbanized areas, otherwise $K = 0$.

4.3.6 McGeehan and Griffiths Model

This is an empirical model obtained as a modification of the plane-earth equation [13]. The path loss (in decibels) is given by

$$L = 120 - 20\log(h_b h_m) + 40\log d + 30\log f + A \tag{4.24}$$

where h_b is the height of the base station antenna, h_m is the height of the mobile antenna, f is the frequency (in megahertz), d is the distance transmitter-receiver (in kilometers), and A is given for various environments as

$$A = \begin{cases} 45 \pm 5 & \text{for older cities with narrow, twisting streets} \\ 55 \pm 5 & \text{for modern cities with long, straight, wide streets} \\ 65 \pm 5 & \text{for suburban areas with some rural areas} \\ 75 \pm 5 & \text{for open areas} \end{cases}$$

4.3.7 Atefi and Parsons Model

This is an empirical model that predicts the median path loss as follows [14]:

$$L = 82 + 26.16\log f + 38\log d - 21.8\log h_b - 0.15\log h_m + L_D \tag{4.25}$$

where h_b is the height of the base station antenna, h_m is the height of the mobile antenna, f is the frequency (in megahertz), d is the distance transmitter-receiver (in kilometers), and L_D accounts for the diffraction loss. This term is evaluated using the Epstein-Peterson method [15].

4.3.8 Sakagami-Kuboi Model

This empirical model requires more detailed information aboaut the environment than the previous models [16]. It estimates the path loss as follows:

$$\begin{aligned} L = {} & 100 - 7.1\log w + 0.023\phi + 1.4\log h_s + 6.1\log H_1 \\ & - \left[24.37 - 3.7\left(\frac{H}{h_{b0}}\right)^2\right]\log h_b \\ & + (43.2 - 3.1\log h_b)\log d + 20\log f + \exp[13(\log f - 3.23)] \end{aligned} \tag{4.26}$$

where w is the width (in meters) of the street where the receiver is located, ϕ is the angle (in degrees) formed by the street axes and the direction of the incident wave, h_s is the height (in meters) of the buildings close to the receiver, H_1 is the average height (in meters) of the buildings close to the reception point, h_b is the height (in meters) of the transmitter antenna with respect to the observer, h_{b0} is the height (in meters) of the transmitter antenna with respect to the ground level, H is the average height (in meters) of the buildings close to the base station, d is the separation between the transmitter and the receiver (in kilometers), and f is the frequency (in megahertz).

The valid ranges of the above parameters are:

$5\text{m} < w < 50\text{m}$, $0° < \phi < 90°$, $5\text{m} < h_s < 80\text{m}$, $5\text{m} < H_1 < 50\text{m}$,
$20\text{m} < h_b < 100\text{m}$, $h_{b0} \geq H$, $0.5\text{km} < d < 10\text{km}$, and
$450 \text{ MHz} < f < 2{,}200 \text{ MHz}$

4.3.9 Ikegami Model

This semideterministic model is suitable for homogeneous urban environments [17]. To derive the equations of the model, GTD and GO techniques (see Sections 2.3 and 2.4) are applied to an ideal city with buildings of uniform height. A NLOS situation is assumed and only two rays are considered (see Figure 4.1): (1) rays diffracted at the last edge before the mobile (dr) and (2) rays reflected at the next building wall (rr).

Under the above assumptions, the path loss (in decibels) is given by:

$$L = 26.65 + 30 \log f + 20 \log d - 10 \log\left(1 + \frac{3}{l_r^2}\right) - 10 \log w \quad (4.27)$$
$$+ 20 \log(h_B - h_r) + 1 \log(\sin\phi)$$

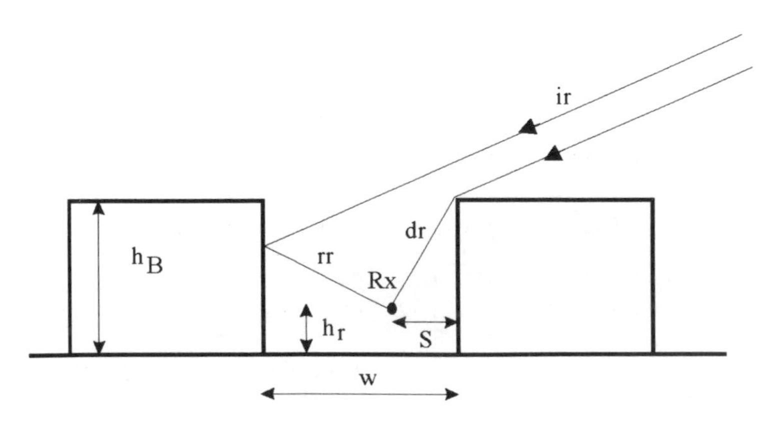

Figure 4.1 Side view of the environment indicating the rays and the parameters considered in the model.

where f is the frequency (in megahertz), d is the separation between the transmitter and the receiver antennas, w is the width of the street where the receiver is placed (Rx), h_B is the building height, h_r is the height of the receiver antenna, ϕ is the angle between the street axes and the directions of the incident rays (ir), and l_r is a parameter that depends on the reflection coefficients of the building faces. A typical value for l_r in the UHF band is 3.2.

This model provides good results when the angle formed by the incident ray (ir) and the horizontal is not low. In the other cases, other mechanisms not considered in the model, can have an important contribution to the field at the receiver.

4.3.10 Walfisch and Bertoni Model

This is also called the diffracting screens model [18]. It is a semideterministic model suitable for homogeneous urban and suburban areas. The propagation equations are derived from an ideal model of a city where the buildings are organized such that they form parallel rows with uniform height and width. Most of the residential areas in North American cities have this structure. The model is valid when there is a NLOS situation between the base station and the receiver. In such cases, the field propagated from the transmitter to the receiver antenna suffers a multiple forward diffraction past rows of buildings. To evaluate the field at the receiver, the rows are modeled as a set of absorbing diffracting screens that are responsible for propagation losses due to the proximity of the buildings. The field at the last rooftop is diffracted and reaches the receiver directly. Also, the field diffracted in the last rooftop is reflected in the next building and then it reaches the receiver. Considering the above mechanism, the path loss is estimated as follows:

$$L = 89.55 + 21 \log f + 38 \log d - 18 \log H + A - 18 \log\left(1 - \frac{d^2}{17H}\right) \tag{4.28}$$

where f is the frequency (in megahertz), d is the distance transmitter-receiver (in kilometers), and H is the average height of the transmitter antenna with respect to the height of the surrounding buildings. The last term accounts for the curvature of the earth, and A is a term that models the influence of the buildings:

$$A = 5 \log\left[\left(\frac{b}{2}\right)^2 + (h_B - h_m)^2\right] - 9 \log b + 20 \log\left\{ \mathrm{tg}^{-1}\left[\frac{2(h_B - h_m)}{b}\right]\right\} \tag{4.29}$$

where h_B is the building height (in meters), h_m is the height of the receiver (in meters), and b is the row spacing (see Figure 4.2). The model requires that the transmitter antenna be above the rooftop level. (Notice that (4.28) breaks down when H is close to zero.)

4.3.11 Xia and Bertoni Model

Recently, the Walfisch and Bertoni model has been improved to allow the transmitter antenna to be below as well as above the rooftops [19–21]. This extended model requires that the distance between the source and the plane of the nearest edge be the same as the distance between the rows of buildings. The path loss (in decibels) at the receiver is calculated as the sum of three terms:

$$L = L_0 + L_1 + L_2 \tag{4.30}$$

where L_0 is the free-space loss (see (4.4)), L_1 is the loss due to the last diffraction down below the rooftop level, and L_2 accounts for the multiple diffracted field along the rooftops at the last edge just before the receiver. The second term can be derived form the GTD/UTD equations of Chapter 2. It is given by

$$L_1 = 10 \log\left[\frac{D^2(\theta)}{\pi k \cos\phi r}\right] \tag{4.31}$$

where, as Figure 4.2 shows, r is the distance from the edge to the receiver, x is the horizontal distance (measured perpendicular to the building walls) from the diffracting edge of the buildings to the receiver (in meters), k is the wave number (in meters^{-1}), ϕ is the angle formed by the street axes and the line

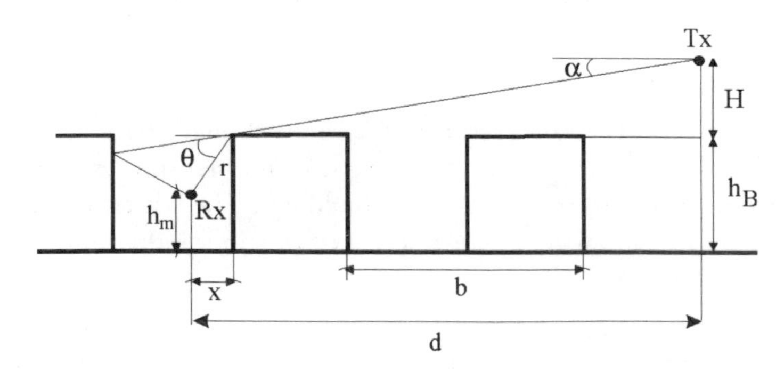

Figure 4.2 Side view of the environment indicating the rays and the parameters considered in the model.

connecting the transmitter and receiver antennas, θ is the angle formed by the diffracted ray and the horizontal, and $D(\theta)$ is the GTD diffraction coefficient given by

$$D(\theta) = \frac{1}{\theta} - \frac{1}{\theta + 2\pi} \tag{4.32}$$

As stated in Section 2.4, when the diffracted ray is in the edge transition zone, (4.31) becomes invalid and the UTD diffraction coefficients of Section 2.4 must be used. This happens for low values of θ.

If the edge is formed by the vertical wall and the rooftop, then

$$\theta = \mathrm{tg}^{-1}\left(\frac{h_B - h_m}{x}\right) \tag{4.33}$$

$$r = \sqrt{x^2 + (h_B - h_m)^2} \tag{4.34}$$

The third term (L_2) is derived from the application of the PO to the multiple diffraction along a set of absorbing screens (see Section 2.9). The resulting expression is given by

$$L_2 = 20 \log Q(g_p) \tag{4.35}$$

where Q is

$$Q(g_p) = 3.502 g_p - 3.327 g_p^2 + 0.962 g_p^3 \tag{4.36}$$

g_p being a dimensionless parameter given by

$$g_p = \mathrm{tg}^{-1}\left(\frac{H}{1000d}\right)\sqrt{\frac{b \cos\phi}{\lambda}} \tag{4.37}$$

where λ is the wavelength. Equation (4.36) is valid for values of g_p in the range $0.01 < g_p < 1.0$, which covers most of the macrocell situations.

This model assumes flat terrain, but irregular terrain effects can be easily incorporated into the model [21].

4.3.12 COST 231-Walfisch-Ikegami Model

This is a semideterministic model suitable for urban environments [10]. It is based on the Walfisch-Bertoni model and the Ikegami model. It includes some empirical corrections to adapt the model to the features of European cities. There are no restrictions on the position of the transmitter antenna with respect to the rooftop level. As in the Walfisch-Bertoni model, the path loss is calculated as the sum of three terms

$$L = L_0 + L_1 + L_2 \tag{4.38}$$

where L_0 is the free-space loss (see (4.4)), L_1 is the loss due to the last diffraction down below rooftop level, and L_2 accounts for the multiple diffraction along the rooftops (except the last diffraction). The second term is given by

$$L_1 = -16.9 - 10\log w + 10\log f + 20\log(h_R - h_m) + L_{11}(\phi) \tag{4.39}$$

where w is the width of the street where the receiver is (in meters), h_R is the average height of the buildings (in meters), h_m is the receiver antenna height (in meters), and

$$L_{11}(\phi) = \begin{cases} -10 + 0.3571\phi & 0 < \phi < 35^\circ \\ 2.5 + 0.075(\phi - 35^\circ) & 35^\circ \le \phi < 55^\circ \\ 4 - 0.1114(\phi - 55^\circ) & 55^\circ \le \phi \le 90^\circ \end{cases} \tag{4.40}$$

where ϕ is the angle formed by the street axes and the line connecting the transmitter and receiver antennas. Finally, the third term

$$L_2 = L_{21} + k_a + k_d \log d + k_f \log f - 9\log b \tag{4.41}$$

where

$$L_{21} = \begin{cases} -18\log(1 + h_B - h_R), & h_B \ge h_R \\ 0, & h_B < h_R \end{cases} \tag{4.42}$$

$$k_a = \begin{cases} 54, & h_B \ge h_R \\ 54 - 0.8(h_B - h_R), & h_B < h_R \wedge d \ge 0.5\text{m} \\ 54 - 0.4d(h_B - h_R), & h_B < h_R \wedge d < 0.5\text{m} \end{cases} \tag{4.43}$$

$$k_d = \begin{cases} 18, & h_B \ge h_R \\ 18 - \dfrac{15(h_B - h_R)}{h_R}, & h_B < h_R \end{cases} \tag{4.44}$$

$$k_f = -4 + k_{f1}\left(\frac{f}{925} - 1\right) \tag{4.45}$$

where h_B is the transmitter antenna height, b is the distance between the centers of adjacent rows of buildings, and k_{f1} is 1.5 in metropolitan centers and 0.7 otherwise. The parameter validity ranges are

800 MHz $\leq f \leq$ 2,000 MHz, 4m $\leq h_B \leq$ 50m, 1m $\leq h_m \leq$ 3m,
0.02 km $\leq d \leq$ 5 km

The preceding equations are valid for a NLOS situation. For a LOS case, the path loss is calculated with an expression similar to the free-space loss equation but with different coefficients that are experimentally obtained from measurements:

$$L = 42.6 + 26 \log d + 20 \log f \tag{4.46}$$

where d must be bigger than 0.02 km.

4.3.13 Other Models Based on the Diffracting Screens Layout

Apart from the models of Xia-Bertoni and COST 231-Walfisch-Ikegami, various models can be obtained from the diffracting screen configuration. These models assume that the field at the receiver is a result of the same mechanisms mentioned in the Walfisch-Bertoni model, but they differ in the expression for the multiple diffraction along the absorbing screens (the L_2 term in the Xia-Bertoni model).

Andersen [22] and Zhang [23] have proposed UTD-based expressions to account for the multiple diffraction (see Section 2.8) whereas Saunders and Bonar [24] derive a PO-based expression (see Section 2.9).

4.4 Microcell Propagation Models

4.4.1 Two-Ray Model

The field at the receiver is calculated considering only the contribution of the direct ray and the ground reflected ray. Therefore, the path loss (in decibels) at the receiver is given by

$$L = 20 \log\left[\left(\frac{\lambda}{4\pi}\right)\left|\frac{\exp(-jkr_1)}{r_1} + \Gamma\frac{\exp(-jkr_2)}{r_2}\right|\right] \quad (4.47)$$

where Γ is the Fresnel coefficient of the ground (see Chapter 2), λ is the wavelength, k is the wave number, and r_1 and r_2 are the direct and reflected path lengths, respectively.

This model is adequate for rural environments with flat ground. But it is also suitable for microcells with low base station antennas where there is an LOS between the transmitter and receiver antennas. In such cases, reflections and diffractions also occur at the walls of the buildings. These contributions result in rapid variations in the simple two-ray model but do not change the overall path loss predicted by the two-ray model [25, 26].

If the path loss given by (4.47) is written as a function of the distance between transmitter d and the receiver, it can be seen that the dependence of L on d can be approximated by two straight sections with different slopes (n_1 and n_2). The break point (also called the turning point) between the sections appears at a distance to the transmitter given by

$$d_b = \frac{4h_t h_r}{\lambda} \quad (4.48)$$

where h_t and h_r are the transmitter and the receiver antenna heights, respectively. The break point distance coincides with the point where the Fresnel ellipse of the direct ray, from the transmitter to the receiver, touches the ground.

Therefore, the path loss can be written as

$$L = \begin{cases} L_1 + 10n_1 \log d & \text{when } 1 < d < d_b \\ L_1 + 10n_2 \log(d/d_b) + 10n_1 \log d_b & \text{when } d > d_b \end{cases} \quad (4.49)$$

with L_1 being the path loss (in decibels) at the reference distance of 1m. For the theoretical two-ray ground reflection model, the values of n_1 and n_2 are 2 and 4, respectively. Measurements in urban microcells at 1800 and 1900 MHz [27] show values between 2.0 and 2.3 for n_1 and between 3.3 and 13.3 for n_2. This approximation is known as the dual-slope model.

A modification on the original model has been proposed by the UIT-R 8/1 group, where three sections are considered instead of two. The predicted path loss is a function of the distance break point exclusively:

$$L = \begin{cases} 40 + 25 \log d, & \text{when } d < \dfrac{d_b}{2} \\ 40 + 25 \log\left(\dfrac{d_b}{2}\right) + 40 \log\left(\dfrac{2d}{d_b}\right), & \text{when } \dfrac{d_b}{2} \le d < 4d_b \\ 40 + 25 \log\left(\dfrac{d_b}{2}\right) + 40 \log(4d_b) + 60 \log\left(\dfrac{d}{4d_b}\right), & \text{when } d \ge 4d_b \end{cases} \tag{4.50}$$

4.4.2 Multiple-Ray Models

Multiple-ray models have been used in urban microcells, under LOS situations, when the Tx and Rx antennas are well below rooftop levels. These models assume the so-called "dielectric canyon" configuration of the street (also called waveguide configuration). The field at the receiver comes from the direct ray between Tx and Rx, rays reflected along the ground, and rays reflected off the vertical planes of the canyon (building walls). An infinite number of multiple reflected rays reach the transmitter so, in practice, these models consider only the most important of those rays. The two-ray model can be viewed as a multiple-ray model in which only two rays are considered. Four-ray and six-ray models have also been proposed. The four-ray model consists of the direct ray, the ground reflected ray, and the two rays singly reflected by the building walls. The six-ray model [25] accounts for the same mechanisms as the four-ray model plus the two doubly reflected rays by the building walls. In such models, all the ray contributions are calculated by using the GTD approach (see Section 2.4).

4.4.3 Multiple-Slit Waveguide Model

The multiple-ray models, when applied to urban environments assume that there are no discontinuities between the buildings along the street. Blaunstein and Levin have proposed a more complete model derived from the multiple-slit waveguide configuration [28, 29]. This model assumes a city configuration formed by two parallel rows of screens (simulating building walls) with randomly distributed slits (gaps between buildings). This model considers the direct field, the multiple GTD-reflections on buildings walls, multiple UTD-diffraction at their corners, and the GTD-reflection from the ground. The resulting equations predict the experimental location of the break point more accurately than the two-ray model. The power decay after the break point is exponential and not a power law as assumed in many break point models. This can explain why the power law coefficients forced/fitted to experiments (after the break point) are often very high. The model can also explain the absence of the break point in some urban conditions observed experimentally.

4.4.4 Uni-Lund Model

This is a microcell model, developed at the University of Lund in Sweden, that is valid for a transmitter antenna below rooftop level [30, 31] (also mentioned in [32]). The model considers two situations: LOS and NLOS. In the first case, the path loss is calculated as

$$L = 10 \log k + \frac{10}{4} \log(l_1^4 + l_2^4) \tag{4.51}$$

with

$$l_1 = d^{n_1},\ l_2 = d^{n_2} d_b^{n_1 - n_2} \tag{4.52}$$

where d is the distance between the Tx and Rx, d_b is the break point distance given by (4.49), and k, n_1 and n_2 are parameters whose values are fixed from the measurements. The resulting values are similar to the values predicted by the dual-slope model, but in this case, the transition between the two sections at the break point is smoother.

In the situation shown in Figure 4.3, the path loss at the NLOS observation points (e.g., the Rx point in the figure) is calculated as the sum of two terms. The first term is the path loss at point O given by the LOS equations. The second term is

$$L = 10[u(d_1) - u(d_2)] \log\left(\frac{d_2}{d_0}\right)^n \frac{\log d - \log d_1}{\log d_2 - \log d_1} + 10u(d_2) \log\left(\frac{d}{d_0}\right)^n \tag{4.53}$$

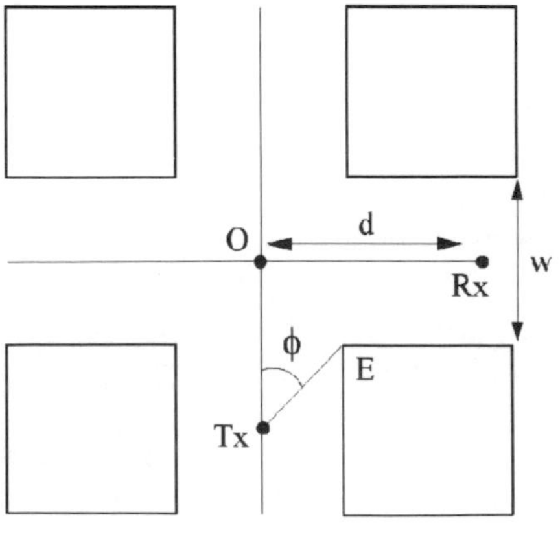

Figure 4.3 Top view of the crossing street scenario.

where $u(x)$ is the unit step function and

$$d_0 = 8.92\phi + 1.7 \tag{4.54}$$

$$d_1 = 10.7\phi + 0.11w + 2.99 \tag{4.55}$$

$$d_2 = 0.31w + 4.9 \tag{4.56}$$

$$n = 2.75 - 1.13\exp(-23.4\phi) \tag{4.57}$$

and w is the street width and ϕ is the angle formed by the street axis of the transmitter and a line joining Tx with the edge (E) of the shadowing building, as is depicted in Figure 4.3.

4.5 Indoor Propagation Models

4.5.1 Log-Distance Path Loss Model

This is an empirical model that estimates the path loss (in decibels) as follows:

$$L = L(d_0) + 10n\log\left(\frac{d}{d_0}\right) + X_\sigma \tag{4.58}$$

where d is the transmitter-receiver separation, $L(d_0)$ is the reference path loss, d_0 is the distance at which path loss is referred to (in meters), n is the path loss exponent which depends on the environment, and X_σ is a normal random variable with a standard deviation of σ (in decibels). The reference path loss can be measured or calculated using the free-space path loss expression. The term X_σ accounts for the environmental clutter. The parameters of the model [$L(d_0)$, n, and σ] depend on the characteristics of the scenario. Several types of indoor environments in LOS and NLOS situations have been analyzed and, for each case, values at different frequencies have been obtained from measurements [33]. At several frequencies between 900 and 4000 MHz, the values of n varied between 1.6 and 3.3 and the values of σ varied between 3.0 and 14.1 dB. Due to its simplicity, this model has been widely used in indoor environments. Also, it has been used in outdoor microcell environments [27].

4.5.2 Attenuation Factor Models

This is an empirical indoor model [34] that predicts the propagation path loss (in decibels) on the same floor or through different floors. The path loss is given by

$$L = L(d_0) + 10n \log\left(\frac{d}{d_0}\right) + \text{FAF} \tag{4.59}$$

where n is the path loss exponent for observers located on the same floor, and FAF represents the floor attenuation factor (in decibels), which accounts for the propagation through different floors. For propagation through multiple floors, when the path loss exponent n_1 is known, an alternative expression is

$$L = L(d_0) + 10n_1 \log\left(\frac{d}{d_0}\right) \tag{4.60}$$

Experimental values of n, n_1, and FAF for various types of buildings can be found in [34]. As an example, the values of the floor attenuation factor (in decibels) varied between 12.9 and 16.2 for transmission through one floor, between 18.7 and 27.5 through two floors, and between 24.4 and 31.6 through three floors. A typical value for n is 2.8. The values of n_1 varied between 4.19 and 5.22. This model is also known as the one-slope model because it assumes that the path loss depends, linearly, on the logarithm of the separation between Tx and Rx.

A modification of the model is proposed in [35] where the path loss is given by

$$L = L(d_0) + 20 \log\left(\frac{d}{d_0}\right) + \alpha d + \text{FAF} \tag{4.61}$$

where α is an attenuation factor determined experimentally for different indoor environments. This model is known as the linear-slope model because the logarithmic path loss depends linearly on the separation between Tx and Rx. Experimental values from [35] at 850, 1700, and 4000 MHz show a variation in the attenuation factor between 0.62 and 0.47 dB/m for a four-story building and between 0.48 and 0.23 dB/m for a two-story building.

4.5.3 Keenan-Motley Model

A more sophisticated method that considers the attenuation through the individual walls and floors is given by [36]

$$L = L_0 + 10n \log d + \sum_{i=1}^{I} k_{fi} L_{fi} + \sum_{j=1}^{J} k_{wj} L_{wj} \tag{4.62}$$

where L_0 is the attenuation at the reference distance (1meter), n is the path loss exponent, d is the distance transmitter-receiver, L_{fi} is the attenuation through floors of the type i, k_{fi} is the number of floors of the type i between the transmitter and the receiver, L_{wj} is the attenuation through walls of type j, and k_{wj} is the number of walls of type j between the transmitter and the receiver. In this model, L_0 and n tend to be the values of the free-space conditions (L_0 = 37, n = 2). Typical values for the attenuation through floors are between 12 and 32 dB [34]. The values of the attenuation through walls depend strongly on the type of partitions used. For typical soft partitions, the attenuation values vary between approximately 1 and 5 dB, and through hard partitions the attenuation can vary between 5 and 20 dB.

4.5.4 Multiple-Wall Model

To fit the measurements better, the Keenan and Motley model has been modified by including a nonlinear function for the number of penetrated floors [32, 37]. The path loss is given by

$$L = L_{FS} + L_C + L_f k_f^{E_f} + \sum_{j=1}^{J} k_{wj} L_{wj} \tag{4.63}$$

where L_{FS} represents the free-space losses between Tx and Rx, L_C is a constant, L_{wj} is the attenuation through walls of type j, k_{wj} is the number of walls of type j between the transmitter and the receiver, k_f is the number of floors between transmitter and receiver, L_f represents the attenuation through adjacent floors, and the exponent E_f in the third term is given by

$$E_f = \frac{k_f + 2}{k_f + 1} - b \tag{4.64}$$

where b is a constant that must be determined empirically. Typical parameter values are L_f = 18.3 dB, I = 2, L_{w1} = 3.4 dB, L_{w2} = 6.9 dB, and b = 0.46 where L_{w1} are the losses through narrow walls (less than 10 cm) and L_{w2} are the losses through wide walls (wider than 10 cm).

References

[1] Lee, W. C. Y., *Mobile Communications Design Fundamentals,* 2nd ed., New York: John Wiley, 1993.

[2] Fleury, B. H., and P. E. Leuthold, "Radiowave Propagation in Mobile Communications: An Overview of European Research," *IEEE Communications Mag.,* Vol. 34, No. 2, Feb. 1996, pp. 70–81.

[3] Allsebrook, K., and J. D. Parsons, "Mobile Radio Propagation in British Cities at Frequencies in the VHF and UHF Bands," *IEEE Proc.,* Vol. 124, No. 2, 1977, pp. 95–102.

[4] Allsebrook, K., and J. D. Parsons, "Mobile Radio Propagation in British Cities at Frequencies in the VHF and UHF Bands," *IEEE Trans. on Vehicular Technology,* Vol. 26, 1997, pp. 313–322.

[5] *Atlas of Radio Wave Propagation Curves for Frequencies Between 30 and 10.000 MHz,* Radio Research Laboratory, Ministry of Postal Services, Tokyo, Japan, 1957, pp. 172–179.

[6] Deslile, G. Y., J. P. Lefevre, M. Lecours, and J. Y. Chourinard, "Propagation Loss Prediction: A Comparative Study With Application to the Mobile Radio Channel," *IEEE Trans. on Vehicular Technology,* Vol. 34, No. 2, 1985, pp. 86–95.

[7] Okumura, Y., E. Ohmori, T. Kawano, and K. Fukuda, "Field Strength and Its Variability in VHF and UHF Land-Mobile Radio Service," *Rev. Electrical Communication Lab.,* Vol. 16, Sep.–Oct. 1968, pp. 825–873.

[8] Hata, M., "Empirical Formula for Propagation Loss in Land Mobile Radio Services," *IEEE Trans. on Vehicular Technology,* Vol. 29, Aug. 1980, pp. 317–325.

[9] Siwiak, K., *Radiowave Propagation and Antennas for Personal Communications,* Norwood, MA: Artech House, 1995.

[10] EURO-COST 231, "Urban Transmission Loss Models for Small-Cell and Micro-Cell Mobile Radio in the 900 and 1800 MHz Bands," Propagation Models Report No. COST231 TD(90) 119, revision 1, Sep. 1991.

[11] Ibrahim, M. F., and J. D. Parsons, "Signal Strength Prediction in Built-Up Areas. Part 1: Median Signal Strength," *IEE Proc.,* Vol. 130, Part F, No. 5, 1983, pp. 377–384.

[12] Parsons, J. D., *The Mobile Radio Propagation Channel,* London: Pentech Press, 1992.

[13] McGeehan, J. P., and J. Griffiths, "Normalised Prediction Chart for Mobile Radio Reception," Proc. 4th International Conf. on Antennas and Propagation 1985, IEE Conference Publication No. 248, pp. 395–399.

[14] Atefi, A., and J. D. Parsons, "Urban Radio Propagation in Mobile Radio Frequency Bands," Proc. Comms 86, Birmingham, UK, IEE Conference Publication No. 262, pp. 13–18.

[15] Epstein, J., and D. W. Peterson, "An Experimental Study of Wave Propagation at 850 MHz," *Proc. IRE,* Vol. 41, No. 5, 1953, pp. 595–611.

[16] Hernando Rábanos, J. M., *Comunicaciones Móviles,* Madrid: De. Centro de Estudios Ramón Areces, 1997 (in Spanish).

[17] Ikegami, F., S.Yoshoida, T. Takeuchi, and M. Umehira, "Propagation Factors Controlling Mean Field Strength on Urban Streets," *IEEE Trans. on Antennas and Propagation,* Vol. 32, Dec. 1984, pp. 822–829.

[18] Walfisch, J., and H. L. Bertoni, "A Theoretical Model of UHF Propagation in Urban Environments," *IEEE Trans. on Antennas and Propagation,* Vol. 36, Oct. 1988, pp. 1788–1796.

[19] Xia, H. H., and H. L. Bertoni, "Diffraction of Cylindrical and Plane Waves by an Array of Absorbing Half Screens," *IEEE Trans. on Antennas and Propagation,* Vol. 40, No. 2, Feb. 1992, pp. 170–177.

[20] Maciel, L. R., H. L. Bertoni, and H. H. Xia, "Unified Approach to Prediction of Propagation Over Buildings for All Ranges of Base Antenna Height," *IEEE Trans. on Vehicular Technology,* Vol. 42, No. 1, Feb. 1993, pp. 41–45.

[21] Bertoni, H., W. Honcharenko, L. R. Maciel, and H. H. Xia, "UHF Propagation Prediction for Wireless Personal Communications," *Proc. IEEE,* Vol. 82, No. 9, Sep. 1994, pp. 1333–1356.

[22] Andersen, J. B., "UTD Multiple-Edge Transition Zone Diffraction," *IEEE Trans. on Antennas and Propagation,* Vol. 45, No. 7, July 1997, pp. 1093–1097.

[23] Zhang, W., "A Wide-Band Propagation Model Based on UTD for Cellular Mobile Radio Communications," *IEEE Trans. on Antennas and Propagation,* Vol. 45, No. 11, Nov. 1997, pp. 1669–1678.

[24] Saunders, S. R., and F. R. Bonar, "Prediction of Mobile Radio Wave Propagation Over Buildings of Irregular Heights and Spacing," *IEEE Trans. on Antennas and Propagation,* Vol. 42, No. 2, Feb. 1994, pp. 137–144.

[25] Rustako, A. J., N. Amitay, G. J. Owens, and R. S. Roman, "Radio Propagation at Microwave Frequencies for Line-of-Sight Microcellular Mobile and Personal Communications," *IEEE Trans. on Vehicular Technology,* Vol. 40, 1991, pp. 203–210.

[26] Xia, H. H., H. L. Bertoni, L. R. Maciel, A. Lindsay-Steward, and R. Rowe, "Radio Propagation Characteristics for Line-of-Sight Microcellular and Personal Communications," *IEEE Trans. on Antennas and Propagation,* Vol. 41, No. 10, 1993, pp. 1439–1447.

[27] Feuerstein, M. J., K. L. Blackard, T. S. Rappaport, S. Y. Seidel, and H. H. Xia, "Path Loss, Delay Spread, and Outage Models as Functions of Antenna Height for Microcellular System Design," *IEEE Trans. on Vehicular Technology,* Vol. 43, No. 3, Aug. 1994, pp. 487–498.

[28] Blaunstein, N., and M. Levin, "VHF/UHF Wave Attenuation in a City With Regularly Spaced Buildings," *Radio Science,* Vol. 31, No. 2, Mar./Apr. 1996, pp. 313–323.

[29] Blaunstein, N., R. Giladi, and M. Levin, "Characteristics' Prediction in Urban and Suburban Environments," *IEEE Trans. on Vehicular Technology,* Vol. 47, No. 1, Feb. 1998, pp. 225–234.

[30] Börjeson, H., C. Bergljung, and L. G. Olson, "Outdoor Microcell Measurements at 1700 MHz," Proc. 41th IEEE Vehicular Technology Conference, 1992.

[31] Börjeson, H., C. Bergljung, P. Karlsson, L. O. Olsson, and S.-O.Ohrvik, "Using a Novel Model for Predicting Propagation Path Loss," COST 231 TD(93) 86, Grimstad, Norway, May 1993.

[32] Cichon, D. J., and T. Kürner, "Propagation Prediction Models," COST 231, Final Report, 1996, pp. 1–83.

[33] Andersen, J. B., T. S. Rappaport, and S. Yoshida, "Propagation Measurements and Models for Wireless Communications Channels," *IEEE Communications Mag.,* Nov. 1994, pp. 42–49.

[34] S. Y. Seidel, T. S. Rappaport. "914 MHz Path Loss Prediction Models for Indoor Wireless Communications in Multifloored Buildings," *IEEE Trans. on Antennas and Propagation,* Vol. 40, No. 2, Feb. 1992, pp. 207–217.

[35] Devasirvatham, D. M. J., C. Banerjee, M. J. Krain, D. A. Rappaport. "Multi-frequency radiowave propagation measurements in the portable radio environment," IEEE International Conference on Communications, 1990, pp. 1334–1340.

[36] Keenan, J. M., A. J. Motley. "Radio coverage in buildings," *British Telecom Technology*, Vol. 8, No. 1, January, 1990, pp. 19–24.

[37] Tornevik, C., J.-E. Berg, F. Lotse. "900 MHz propagation measurements and path loss models for different indoor environments," *Proc. IEEE* VTC'93, New Jersey, USA, 1993.

5

Estimation of Radio Channel Parameters From Results of GTD/UTD Models

5.1 Introduction

In previous chapters, we presented several models for the computation of the power coverage in different scenarios. Most of these models are empirical and they give primarily statistical estimations for some important parameters of the received field level: expected value, variance, probability density function (PDF), autocorrelation function of the field level along a short trajectory, etc. These radio channel parameters are required by the radio system designer.

In this chapter we address the computation of some of the most important radio channel parameters from results obtained from deterministic propagation models, in particular from the GTD/UTD model described in Chapter 2.

We will start by computing the baseband response to the impulse function. The bandwidth of the channel and the possible intersymbol interference (ISI) are determined by the bandwidth of this response. Also, the amount of difficulty a channel equalizer experiences in trying to alleviate the ISI problem depends on the shape of the baseband impulse response. For all of these reasons it is worth dedicating some effort to obtaining the baseband impulse response. To do that, we obtain in Section 5.2, the transfer function of the radio channel from the GTD/UTD expressions. The transfer function is rewritten in one useful way for its treatment as a passband signal. The formulation based on the complex envelope function and the theory that allows transformation of a passband signal problem into a baseband problem are reviewed in Section

5.3. This theory is applied in the following section to the transfer function derived from the GTD/UTD expressions. The ideal and approximate baseband expressions of the impulse response are given at the end of Section 5.4.

Given the complexity of the typical wireless scenarios, it is possible to find hundreds of rays coupling the transmitter and the receiver: direct ray, reflected rays, diffracted rays, double-diffracted rays, etc. Each one of these coupling mechanisms suffers phase delays that cause considerable phase changes if the Tx or Rx stations move a fraction of a wavelength or if the GTD/UTD computations of the exact allocation of the scatterers (walls, edges, etc.) are mistaken by some decimeters. These facts result in a field level that looks like a random signal that must be treated in a statistical way, because it does not make sense to calculate, at one particular point, the deterministic value of the amplitude and phase of the received signal. Depending on the relationship among the amplitudes of the different rays arriving at the receiver, we use one or another statistical model.

When we have a very reduced number of significant rays (e.g., the Rx is close to the Tx in an area where the field is mainly due to the direct ray and two or three reflected rays), there is no particular statistical distribution that can be used to describe the field behavior. In these cases a dedicated statistical analysis of the field distribution can be performed by means of the characteristic functions we can associate with each ray. Lebherz, et al. [1], and Kürner, et al. [2], show how to apply the characteristic functions theory to the complex vectors obtained from GTD/UTD formulations.

However, we are often in areas where the relative position between the Tx and Rx stations is such that the received signal is formed by a large amount of coupling mechanisms. In these cases there are a couple of statistical distributions that describe the received field behavior reasonably well: the Rayleigh and the Rice distributions. The Rayleigh distribution is applicable when all the coupling mechanisms have a similar amplitude, and the second one is useful when there is one predominant ray over a large number of rays with similar amplitudes [3–6]. Section 5.5 shows how to apply these distributions and how to obtain estimations of some important first-order and second-order statistical parameters of the radio channel.

5.2 Transfer Function Computation Using GTD/UTD Deterministic Models

Until now, we have shown in previous chapters how to compute, at an arbitrary observation point r_0, the field due to a source in point r_s. Harmonic time variation, $\exp(j\omega_0 t)$ has been assumed and we also assume all the GTD/UTD premises follow. Therefore, the time harmonic field at the observation point is given by

$$E_T(r_o,\ \omega_o,\ t) = \left[\sum_i E_i(r_o,\ \omega_o)\right] \exp(j\omega_o t) \tag{5.1}$$

where $E_i(r_o,\ \omega_o)$ is the field of any one of the possible coupling mechanisms between the source and the observation points [7]. The dependence of the field (modulus and phase) on the frequency ω_o has been explicitly shown in (5.1). For an arbitrary term on the right-hand side of (5.1) this dependence can be developed as follows:

$$E_i(r_o,\ \omega_o) = E_{io}(r_o,\ \omega_o) \exp\left(-j\omega_o \frac{d_i}{c}\right) \tag{5.2}$$

where c is the speed of light, d_i is the length of the i path distance, and $E_{io}(r_o,\ \omega_o)$ is a complex function of ω_o that depends on the antenna gains and on the geometry and also includes the coefficient of the coupling mechanism: reflection coefficient, diffraction coefficient, etc. We can assume that $E_{io}(r_o,\ \omega_o)$ varies slowly in the proximity of the carrier frequency of the $E_{io}(r_o,\ \omega_o)$ transmitted signal ω_o, so that:

$$E_i(r_o,\ \omega_o) = |E_{io}(r_o,\ \omega_c)| \exp[-j\Phi_{io} - j\tau_{io}(\omega_o - \omega_c)] \tag{5.3}$$

where Φ_{io} and τ_{io} are the phase and slope of the phase (group delay) of $E_i(r_o,\ \omega_o)$ at ω_c

$$\Phi_{io} = -\text{phase}[E_{io}(r_o,\ \omega_c)] + \omega_c \tau_i \tag{5.4}$$

$$\tau_{io} = -\frac{d}{d\omega}\text{phase}[E_{io}(r_o,\ \omega)]\bigg|_{\omega=\omega_c} + \tau_i \tag{5.5}$$

where the parameter τ_i is the time delay of the propagation of the field along the i path:

$$\tau_i = \frac{d_i}{c} \tag{5.6}$$

From (5.4) we can state that the phase of the field at ω_o has two components: one component that depends on the phase of $E_{io}(r_o,\ \omega_o)$ and another component that depends on the phase delay due to the propagation of the field. Both components of the phase are important. The slope of the phase, given in (5.5), also has two components. However, in most cases the slope can be approximated very accurately by the propagation time delay, because we can presume that the phase slope due to the antennas' gains or reflection or diffraction coefficients is quite small, therefore,

$$\tau_{io} \cong \tau_i \tag{5.7}$$

Expressions (5.1) and (5.2) are the transfer functions of the radio channel if we consider this channel to be a linear and time invariant system (LTI) [8]. These functions can also be used to analyze the received field when the transmitted field supports a more complex time variation than the time harmonic variation $\exp(j\omega_0 t)$. In later sections, we will see how to study the case of the transmission of a passband signal.

5.3 Impulse Response Model for Passband Systems

In most of the wireless applications, the transmitted signals can be considered to be passband signals, which means that in the spectral domain, power is centered around a carrier frequency ω_c. In addition, we can assume that the spectrum of these signals is negligible for frequencies further away than W from ω_c. A passband signal, $x(t)$ can be represented as follows [9]:

$$x(t) = A(t)\cos[\omega_c t + \phi(t)] \tag{5.8}$$

where $A(t)$ and $\phi(t)$ are the envelope and phase, respectively, of a carrier at frequency ω_c. The passband signal can be developed in terms of the in-phase and quadrature components:

$$x(t) = x_1(t)\cos(\omega_c t) - x_Q(t)\sin(\omega_c t) \tag{5.9}$$

where the in-phase and quadrature components are defined by

$$x_1(t) = A(t)\cos[\phi(t)] \tag{5.10a}$$

$$x_Q(t) = A(t)\sin[\phi(t)] \tag{5.10b}$$

We can define the complex envelope of $x(t)$ from its in-phase and quadrature components as follows:

$$x_c(t) = x_I(t) + jx_Q(t) \tag{5.11}$$

Note that it is quite straightforward to understand that the passband signal can be obtained from the complex envelope by

$$x(t) = \mathrm{Re}[x_c(t)\ \exp(\omega_c t)] \tag{5.12}$$

In a similar way, the impulse response of a passband system $h(t)$ can be developed from its in-phase and quadrature components:

$$h(t) = h_I(t)\cos(\omega_c t) - h_Q(t)\sin(\omega_c t) \tag{5.13}$$

The complex envelope of $h(t)$ is given by:

$$h_c(t) = h_I(t) + jh_Q(t) \tag{5.14}$$

and the passband impulse response can be recovered from the complex envelope by using

$$h(t) = \mathrm{Re}[h_c(t)\ \exp(\omega_c t)] \tag{5.15}$$

Assuming that the passband system is an LTI, its response to the input signal $x(t)$ is also a passband signal, which can be obtained from the convolution operation

$$y(t) = x(t) * h(t) \tag{5.16}$$

Figure 5.1 shows a block diagram of an LTI system.

Moreover, the response signal can be obtained from its complex envelope

$$y(t) = \mathrm{Re}[y_c(t)\ \exp(\omega_c t)] \tag{5.17}$$

x(t)
h(t)
y(t)

Figure 5.1 Passband problem with an LTI system. The output signal $y(t)$ can be obtained from the convolution of the input signal $x(t)$ with the impulse response of the system $h(t)$. If the input signal is narrowband, then the output signal is also narrowband, regardless of whether $h(t)$ is band limited or not. Therefore, we can consider the LTI system to be band limited if the input is band limited.

where the complex envelope of the response can be defined by the convolution between the complex envelopes of the input signal and of the input response [9]:

$$y_c(t) = \frac{1}{2} x_c(t) * h_c(t) \tag{5.18}$$

Figure 5.2 presents a block diagram of an LTI system whose response is defined by this last equation. Note that the LTI system of Figure 5.2 is a baseband (low-frequency) system equivalent to the system of Figure 5.1. Of course, it is easier to perform a signal analysis, either numerically or by means of a hardware implementation, of a low-frequency system than of a passband system.

Developing (5.18) on in-phase and quadrature components, we can derive the following expressions:

$$y_1(t) = \frac{1}{2}[x_1(t) * h_1(t) - x_Q(t) * h_Q(t)] \tag{5.19a}$$

$$y_Q(t) = \frac{1}{2}[x_1(t) * h_Q(t) - x_Q(t) * h_1(t)] \tag{5.19b}$$

where the complex envelope of the response is defined by

$$y_c(t) = y_1(t) + jy_Q(t) \tag{5.20}$$

To illustrate how to go from the passband problem to the equivalent baseband problems, let's consider the problem indicated in Figure 5.3, where a baseband signal $s(t)$ is modulated in amplitude

$$x(t) = s(t) \cos(\omega_c t) \tag{5.21}$$

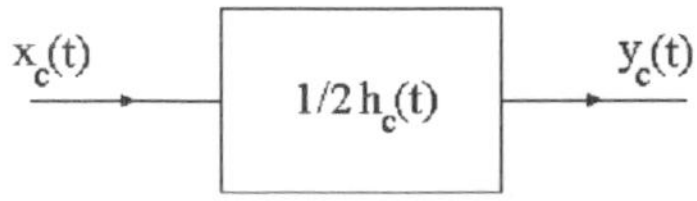

Figure 5.2 Equivalent baseband problem defined using a baseband LTI system. The complex envelope of the output signal, $y_c(t)$, is half of the convolution between the complex envelopes of the input, $x_c(t)$ and the impulse response $h_c(t)$. The three signals in the figure are baseband signals.

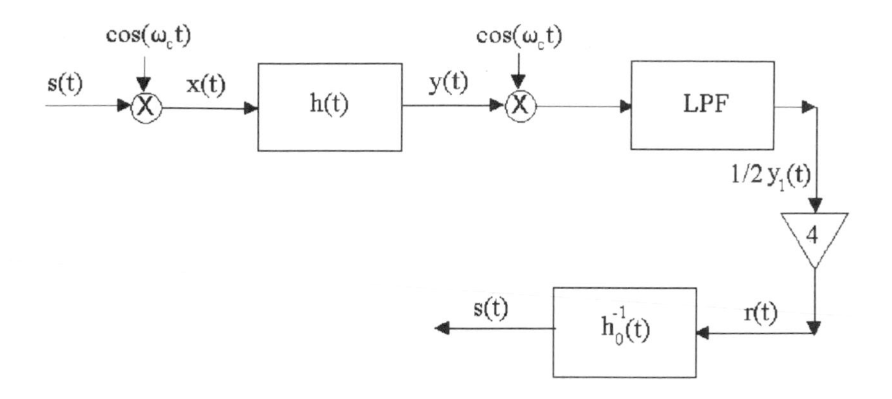

Figure 5.3 Example of a passband transmission: a baseband signal, $s(t)$, is amplitude modulated to give $x(t)$, then transmitted by a channel with impulse response $h(t)$. The received signal $y(t)$ is demodulated to obtain the baseband signal $r(t)$ that must be passed through an equalizer filter to recover the original baseband signal $s(t)$.

The resulting signal is a passband signal with the following in-phase and quadrature components:

$$x_1(t) = s(t) \tag{5.22a}$$

$$x_Q(t) = 0 \tag{5.22b}$$

The passband signal is transmitted through a channel with an impulse response

$$h(t) = h_o(t)\cos(\omega_c t) \tag{5.23}$$

The in-phase and quadrature components of the impulse response are defined by

$$h_1(t) = h_o(t) \tag{5.24a}$$

$$h_Q(t) = 0 \tag{5.24b}$$

To work with an equivalent baseband system, we obtain the complex envelopes of $x(t)$ and $h(t)$:

$$x_c(t) = s(t) \tag{5.25}$$

$$h_c(t) = h_o(t) \tag{5.26}$$

The complex envelope of the output signal will be

$$y_c(t) = \frac{1}{2}x_c(t) * h_c(t) = \frac{1}{2}s(t) * h_o(t) \tag{5.27}$$

whose in-phase and quadrature components are given by

$$y_1(t) = \frac{1}{2}s(t) * h_o(t) \tag{5.28a}$$

$$y_Q(t) = 0 \tag{5.28b}$$

and the received passband signal is obtained from the complex envelope $y_c(t)$ as follows

$$y(t) = \mathrm{Re}[y_c(t) \exp(\omega_c t)] = \left[\frac{1}{2}s(t) * h_o(t)\right] \cos(\omega_c t) \tag{5.29}$$

This signal can be demodulated, filtered with a low-pass filter, and amplified to obtain the received band signal, $r(t)$. This signal is a distorted replica of the original transmitted baseband signal, $s(t)$. We can recover the actual shape of this signal by using the equalizing filters indicated in the last of the blocks of Figure 5.3, which have as an impulse response, $h_o^{-1}(t)$.

The analysis of the system of Figure 5.3 using the baseband envelope functions gives us at least two important advantages. One is that we can analyze the transmission channel and, if necessary, do a computer simulation of the system using low band signals, instead of the very rapidly varying signals $[x(t), y(t), h(t)]$ that the system actually supports. The other advantage is that we can accurately recover the shape of the original transmitted signal by implementing a filter in the baseband, instead of in the radio-frequency band (assuming that the channel in Figure 5.3 is the radio-frequency channel).

Summarizing, we have introduced advantages for finding and working with the envelope of the impulse response. Moreover, many times it is only necessary to find the modulus of the envelope of the impulse function.

5.4 Computation of the Baseband Impulse Functions From the Transfer Functions

The continuous Fourier transform of the complex envelope of the impulse response of a passband LTI system can be written in terms of its in-phase and quadrature components, $H_I(\omega)$ and $H_Q(\omega)$, respectively, as follows:

$$H_c(\omega) = H_I(\omega) + jH_Q(\omega) \tag{5.30}$$

Considering the relationship between the impulse function and its complex envelope [9], (5.13)–(5.15), the transfer function of the system can be developed in terms of shifted replicas of $H_I(\omega)$ and $H_Q(\omega)$:

$$H(\omega) = \begin{cases} \frac{1}{2}H_I(\omega - \omega_c) + j\frac{1}{2}H_Q(\omega - \omega_c) & \text{for } \omega \geq 0 \\ \frac{1}{2}H_I(\omega + \omega_c) - j\frac{1}{2}H_Q(\omega + \omega_c) & \text{for } \omega \leq 0 \end{cases} \tag{5.31}$$

Alternatively, we can find the following inverse relations between the transfer function and the in-phase and quadrature spectral components of the complex envelope of the impulse function:

$$H_I(\omega) = H^+(\omega + \omega_c) + H^-(\omega - \omega_c) \tag{5.32a}$$

$$H_Q(\omega) = -jH^+(\omega + \omega_c) + jH^-(\omega - \omega_c) \tag{5.32b}$$

where$H^+(\omega)$ and $H^-(\omega)$ are the right and left parts of the transfer function that can be defined with the help of the step function $u(\omega)$:

$$H^+(\omega) = H(\omega)u(\omega) \tag{5.33a}$$

$$H^-(\omega) = H(\omega)u(-\omega) \tag{5.33b}$$

The transfer function is a complex magnitude that can be developed in terms of its modulus and phase:

$$H(\omega) = |H(\omega)| \exp[-j\Phi(\omega)] \tag{5.34}$$

Let's assume that we are dealing with a transfer function that varies very slowly around a carrier frequency, ω_c. In this case, we can approach the transfer function around this frequency by one function constant in modulus and with a linear variation of the phase

$$H(\omega) = |H(\omega_c)| \exp[-j\Phi(\omega_c)] \exp[-j(\omega - \omega_c)\tau_c] \tag{5.35}$$

where the phase has been approximated by the first two terms of its Taylor's series:

$$\Phi(\omega) = \Phi(\omega_c) + (\omega - \omega_c)\tau_c \tag{5.36}$$

$$\tau_c = \left.\frac{d\Phi}{d\omega}\right|_{\omega=\omega_c} \tag{5.37}$$

Now using (5.32) we can compute the in-phase and quadrature components of this approximation for the transfer function:

$$H_I(\omega) = 2|H(\omega_c)| \cos[\Phi(\omega_c)] \exp(-j\omega\tau_c) \tag{5.38a}$$

$$H_Q(\omega) = -2|H(\omega_c)| \sin[\Phi(\omega_c)] \exp(-j\omega\tau_c) \tag{5.38b}$$

and using the inverse Fourier transform, we obtain the time-domain components of the complex envelope of the impulse response:

$$h_I(t) = 2|H(\omega_c)| \cos[\Phi(\omega_c)]\, \delta(t - \tau_c) \tag{5.39a}$$

$$h_Q(t) = -2|H(\omega_c)| \sin[\Phi(\omega_c)]\, \delta(t - \tau_c) \tag{5.39b}$$

and the modulus of the complex envelope will be

$$|h_c(t)| = 2|H(\omega_c)|\delta(t - \tau_c) \tag{5.40}$$

Examining expressions (5.39a,b) we can see that both components are delayed impulses, with a time delay defined by the slope around the carrier frequency of the transfer function, (5.37).

In all the development from (5.35) to (5.38), we are assuming that the signals to be analyzed are limited to a frequency band narrow enough such that we can consider applicable the premises that allowed the approximations of (5.35): The modulus of the transfer function is constant and its phase has a linear variation. Of course, the impulse function is not a narrow-banded signal and, rigorously speaking, expressions (5.39) are not completely correct because they have been obtained from assuming that the spectral components of the complex envelope are given by (5.34) for all the frequencies. However, the impulse response given by (5.35) can be used if we are dealing with signals whose bandwidth is small enough that we can consider (5.34) applicable in all the bands of these signals. For instance, if we are using a narrow window function $w(t)$ to represent an "approximate" baseband impulse function (this is the case when we are measuring the channel impulse response), then

$$\delta'(t) = w(t) \cos(\omega_c t) \tag{5.41}$$

and if this window has a bandwidth small enough to allow the use of (5.39), we can derive the components of the envelope of the "approximate" impulse response as follows:

$$h'_I(t) = 2|H(\omega_c)| \cos[\Phi(\omega_c)]\, w(t - \tau_c) \tag{5.42a}$$

$$h'_Q(t) = -2|H(\omega_c)| \sin[\Phi(\omega_c)]\, w(t - \tau_c) \tag{5.42b}$$

and the modulus of the "approximate" complex envelope will be

$$|h'_c(t)| = 2|H(\omega_c)|\, w(t - \tau_c) \tag{5.43}$$

Now we are very well placed to derive the impulse response from the GTD/UTD deterministic model. In particular, if we examine the approximation around the carrier frequency of the response from a coupling mechanism of a time-harmonic excitation, given in (5.3), we can obtain the following transfer function for this coupling mechanism:

$$E_i(r_o, \omega) = |E_{io}(r_o, \omega_c)| \exp(-j\Phi_{io}) \exp[-j\tau_i(\omega - \omega_c)] \tag{5.44}$$

where we must remember that Φ_{io} is the phase of $E_i(r_o, \omega)$ at ω_c, τ_i is the time delay of propagation of the field along the i path, and $E_{io}(r_o, \omega_c)$ is a complex parameter that depends on the antenna gains and on the geometry and also includes the coefficient of the coupling mechanism: reflection coefficient, diffraction coefficient, etc.

For the i-coupling mechanism, the in-phase and quadrature components of the envelope of an "approximate" passband impulse function will be

$$h'_{Ii}(t) = 2|E_{io}(r_o, \omega_c)| \cos(\Phi_{io})\, w(t - \tau_i) \tag{5.45a}$$

$$h'_{Qi}(t) = -2|E_{io}(r_o, \omega_c)| \sin(\Phi_{io})\, w(t - \tau_i) \tag{5.45b}$$

and the modulus of the envelope will be

$$|h'_{ci}(t)| = 2|E_{io}(r_o, \omega_c)|\, w(t - \tau_i) \tag{5.46}$$

The transfer function for the total field can be derived considering (5.1)–(5.3):

$$E_T(r_o, \omega) = \sum_i E_i(r_o, \omega) = \sum_i |E_{io}(r_o, \omega_c)| \exp(-j\Phi_{io}) \exp[-j\tau_i(\omega - \omega_c)] \tag{5.47}$$

and by applying the superposition principle, we can obtain the baseband response (in-phase and quadrature components) of an "approximate" passband impulse:

$$h'_{IT}(t) = 2\sum_i |E_{io}(r_o, \omega_c)| \cos(\Phi_{io})\, w(t - \tau_i) \tag{5.48a}$$

$$h'_{QT}(t) = -2\sum_i |E_{io}(r_o, \omega_c)| \sin(\Phi_{io})\, w(t - \tau_i) \tag{5.48b}$$

The amplitude of the "approximate" impulse response is a long and relatively complicated expression because, due to the nonzero duration of the approximate functions, the impulse responses to the different coupling mechanisms overlap. In particular, we have for the "approximate" complex response:

$$\begin{aligned} |h'_{cT}(t)| &= 2\sqrt{h'^2_{IT}(t) + h'^2_{QT}(t)} \\ &= 2\sqrt{\sum_i \sum_j |E_{io}(r_o, \omega_c)|\,|E_{jo}(r_o, \omega_c)| \cos(\Phi_{io} - \Phi_{jo})\, w(t - \tau_i)\, w(t - \tau_j)} \end{aligned} \tag{5.49}$$

This expression can be simplified if the time-broad of the approximate impulse is smaller than the differences in the time delays, $(\tau_i - \tau_j)$, of any pair of coupling mechanisms. Of course, this happens for the ideal impulse, for which the modulus of the complex envelope is given by

$$|h_{cT}(t)| = 2\sum_i |E_{jo}(r_o, \omega_c)|\, \delta(t - \tau_i) \tag{5.50}$$

Now we have obtained (5.47) and (5.49), which describe the baseband impulse response of the channel in the spectral and time domains, respectively. These equations relate the impulse response to the UTD model. In real life, the parameters of this model, like the amplitude, the phase Φ_{io}, and the time delay τ_i, are not known in a deterministic way. For instance, it is not possible to know the exact coordinates of the walls of the geometrical model; instead, we usually know these coordinates with random error that is often greater than the wavelength, and therefore the amplitude, phase, and time delay of the reflected or diffracted fields should be considered to be random variables. By considering the PDF of the parameters in (5.47) and (5.49), one could obtain a statistical description of the impulse response. In any case, the reader is referred to [10, 11] for the statistical analyses of some of the parameters of the impulse response.

5.5 Channel Statistic Analysis

The field time variation at a point r_o, given by (5.1), can be rewritten in a shorter form as follows:

$$E_T(r_o, \omega_o, t) = \sum_{i=1}^{N} A_i \cos(\omega_o t - \Phi_i) \tag{5.51}$$

where A_i is a positive real number and it represents the amplitude of the *i*-coupling mechanism:

$$A_i = |E_i(r_o, \omega_o)| \tag{5.52}$$

The parameter Φ_i of (5.51) is the phase of this coupling mechanism, defined in (5.4). We can rewrite this phase here as follows:

$$\Phi_i = \Phi_{io} = \omega_o \tau_i - \text{phase}[E_{io}(r_o, \omega_o)] \tag{5.53}$$

In a complex environment the GTD/UTD model gives a large number of coupling mechanisms, *N*, and the parameters A_i and Φ_i can be considered to be statistically independent random variables (RV). The phase Φ_i is an RV uniformly distributed in the interval $(0,2\pi)$. The amplitude Ai is an RV whose mean square value is

$$E\{A_i^2\} = W_i \tag{5.54}$$

where W_i is proportional to the average power density of the *i*-coupling mechanism [6].

The field variation at a point r_o can be expressed in terms of in-phase and quadrature components

$$E_T(r_o, \omega_o, t) = E_I \cos(\omega_o t) - E_Q \sin(\omega_o t) \tag{5.55}$$

where

$$E_I = \sum_{i=1}^{N} A_i \cos(\Phi_i) \tag{5.56a}$$

$$E_Q = \sum_{i=1}^{N} A_i \sin(\Phi_i) \tag{5.56b}$$

Assuming that N is large enough, we can then apply the central limit theorem to the in-phase and quadrature components, causing these components to be independent Gaussian RVs, that can therefore be completely described by their means and their variances [12, 13]. In this case, the means are always zero for both the in-phase and quadrature components. The variance is the same for the two components:

$$\sigma^2 = E\{E_I^2\} = E\{E_Q^2\} = \sum_{i=1}^{N} E\{A_I^2\} = \sum_{i=1}^{N} W_i \tag{5.57}$$

The envelope of the total field (the amplitude of its complex envelope) is given by

$$e = \sqrt{E_I^2 + E_Q^2} \tag{5.58}$$

and the phase by

$$\phi = \mathrm{tg}^{-1}\left(\frac{E_Q}{E_I}\right) \tag{5.59}$$

We are going to distinguish two cases in the analysis of the statistical behavior of the envelope and phase. One situation appears when all the coupling mechanisms that give the total field are, more or less, of the same magnitude. This type of situation can occur, for instance, when the observation point, r_o is in the NLOS of the transmitter. In this case, the statistic of the envelope is described by the Rayleigh density function [14]. The other situation is when one coupling mechanism is substantially greater than the rest of the field components. This last situation frequently appears for points at the LOS of the transmitter and now the statistics of the envelope follow the Rice density function [15]:

5.5.1 Rayleigh's Case

As shown in [14], when all the coupling mechanisms have a similar level, the PDF of the envelope is given by the Rayleigh density function

$$p(e) = \frac{e}{\sigma^2} \exp\left(-\frac{e^2}{2\sigma^2}\right) \tag{5.60}$$

and the CDF (cumulative distribution function)

$$P(e \leq R) = 1 - \exp\left(-\frac{R^2}{2\sigma^2}\right) \tag{5.61}$$

where σ is as defined in (5.57).

From these distribution functions it is easy to obtain the first and second moments, which can all be put in terms of only the parameter σ. For the mean of the envelope we have

$$\bar{e} = E\{e\} = \sqrt{\frac{\pi}{2}}\sigma \tag{5.62}$$

for the mean-square

$$E\{e^2\} = 2\sigma^2 \tag{5.63}$$

for the variance

$$\sigma_e^2 = E\{e^2\} - \bar{e}^2 = \frac{4 - \pi}{2}\sigma^2 \tag{5.64}$$

and for the median

$$e_M = \sqrt{2 \ln 2}\ \sigma \tag{5.65}$$

Regarding the PDF of the phase of the complex envelope, this follows a uniform PDF, as shown in [14]:

$$p(\phi) = \frac{1}{2\pi} \tag{5.66}$$

Another very important parameter of the statistical behavior of the field is its autocorrelation of the amplitude of the envelope when we move in the neighborhood of r_o, along a line. If s and l are values of the distance measured along this line, then the autocorrelation function is defined by

$$C_r(l) = E\{e(s)e(s + l)\} \tag{5.67}$$

As shown in [5], the autocorrelation can be developed in terms of the wavelength as follows:

$$C_r(l) \cong \frac{\pi}{8a(0)} a^2(l, \lambda) \quad (5.68)$$

where function $a(l, \lambda)$ is expressed in terms of the zero-order Bessel function [12]:

$$a(l, \lambda) = \frac{\sigma^2}{2} J_o\left(\frac{2\pi l}{\lambda}\right) \quad (5.69)$$

From the previously described first-order statistical parameters, it is now possible to obtain second-order statistical parameters [6]. For instance, the expected number of times N_R (also called level crossing rate (LCR)) that the envelope crosses a given level R in a line 1m in length is given by

$$N_R = \sqrt{\frac{\pi}{\sigma^2}} \frac{R}{\lambda} \exp\left(-\frac{R^2}{\sigma^2}\right) \quad (5.70)$$

This expression is usually written in terms of the ratio ρ between the reference level R and the median e_M of the envelope amplitude

$$N_R = \sqrt{2\pi} \frac{\rho}{\lambda} \exp(-\rho^2) \quad (5.71)$$

where

$$\rho = \frac{R}{e_M} \quad (5.72)$$

Closely related to N_R, is the average fade length (AFD), which is the expected value of the length of the segments in which the field strength is below the reference level R. In terms of wavelengths, the AFD is given by L_R

$$L_R = \frac{\lambda}{\rho} \frac{\exp(\rho^2) - 1}{\sqrt{2\pi}} \quad (5.73)$$

5.5.2 Rice's Case

Now we have a ray with a field level greater than the level of the rest of the rays. This was studied in detail by Rice in one of his classical papers [15].

Calling e_p the envelope (level) of the predominant coupling mechanism, the PDF of the total received field level is given by the so-called Rice distribution:

$$p(e) = \frac{e}{\sigma^2} \exp\left(-\frac{e^2 + e_p^2}{2\sigma^2}\right) I_o\left(\frac{ee_p}{\sigma^2}\right) \tag{5.74}$$

where I_o represents the modified Bessel function of the first kind and zero order. Most often this PDF is represented using the parameter, defined as

$$K = 10 \log\left(\frac{e_p^2}{2\sigma^2}\right) \quad \text{(dB)} \tag{5.75}$$

$$p(e) = \frac{2e10^{K/10}}{e_p^2} \exp\left[-\frac{e10^{K/10}}{e_p^2}(e^2 + e_p^2)\right] I_o\left(\frac{2e10^{K/10}}{e_p}\right) \tag{5.76}$$

and the resulting expression for the PDF of the phase of the complex envelope

$$p(\phi) = \frac{1}{2\pi} \exp\left(-\frac{e_p^2}{2\sigma^2}\right)\left\{1 + \sqrt{\frac{\pi}{2}} \frac{e_p \cos(\phi)}{\sigma} \exp\left[\frac{e^2 \cos^2(\phi)}{2\sigma^2}\right]\right\} \times \left\{1 + \operatorname{erf}\left[\frac{e_p \cos(\phi)}{\sigma\sqrt{2}}\right]\right\} \tag{5.77}$$

where phase zero corresponds to the phase of the predominant coupling mechanism.

Finally, the autocorrelation function is defined by

$$C_r(l) \cong \frac{\pi}{8} a(0) J_o^2\left(2\pi\frac{l}{\lambda}\right) \tag{5.78}$$

References

[1] Lebherz, M., W. Wiesbeck, W. Krank, "A Versatile Wave Propagation Model for the VHF/UHF Range Considering Three-Dimensional Terrain," *IEEE Trans. on Antennas and Propagation,* Vol. 40, No. 10, Oct. 1992, pp. 1121–1131.

[2] Kürner, T., D. J. Cichon, and W. Wiesbeck, "Evaluation and Verification of the VHF/UHF Propagation Channel Based on a 3-D Wave Propagation Model," *IEEE Trans. on Antennas and Propagation,* Vol. 44, No. 3, Mar. 1996, pp. 393–404.

[3] Clarke, R. H., "A Statistical Theory of Mobile Radio Reception," *Bell Syst. Tech. J.*, Vol. 44, 1968, pp. 957–1000.

[4] Gans, M. M., "A Power-Spectral Theory of Propagation in the Mobile Radio Environments," *IEEE Trans. on Vehicular Technology*, Vol. 21, No. 1, 1972, pp. 27–38.

[5] Jakes, W. C., *Microwaves Mobile Communications*, New York: John Wiley, 1974.

[6] Parsons, J. D., *The Mobile Radio Propagation Channel*, London: Pentech Press, 1992.

[7] Kanatas, A. G., I. D. Kountoris, G. B. Kostaras, and Ph. Constantinou, "A UTD Propagation Model in Urban Microcellular Environments," *IEEE Trans. on Vehicular Technology*, Vol. 46, No. 1, Feb. 1997, pp. 185–193.

[8] Oppenheim, A. V., A. S. Willsky, and H. Nawab, *Signals & Systems*, London: Prentice Hall, 1997.

[9] Haykin, S., *Communication Systems*, New York: John Wiley, 1994.

[10] Bach Andersen, J., S. L. Lauritzen, and C. Thommesen, "Distributions of Phase Derivatives in Mobile Communications," *IEE Proc.*, Part H., Vol. 37, No. 4, Aug. 1990, pp. 197–201.

[11] Crohn, I., E. Bonek, and H. Weinrichter, "A Modified Scatterer Distribution Model for Fast Fading Mobile Communication Channel," *Electronic Letters*, Vol. 26, No. 19, Sep. 1990, pp. 1575–1576.

[12] Davenport, W. B., and W. L. Root, *An Introduction to the Theory of Random Signals and Noise*, New York: McGraw-Hill, 1958.

[13] Papoulis, A., *Probability, Random Variables and Stochastic Processes*, New York: McGraw-Hill, 1991.

[14] Rice, S. O., "Mathematical Analysis of Random Noise," *Bell Syst. Tech. J.*, Vol. 23, 1948, pp. 292–332.

[15] Rice, S. O., "Statistical Properties of a Sine Wave Plus Random Noise," *Bell Syst. Tech. J.*, Vol. 27, 1948, pp. 109–157.

6

Microcellular Design and Channel Allocation

6.1 Introduction

In this chapter we present some techniques for the cell design and channel planning of microcell systems. We start with the definition of the terminology of microcellular design that we will use in the rest of the chapter. After that, a review of the historical evolution of the techniques for the design of macrocell systems [1, 2] will be outlined in order to state the similarities and differences that appear for macro- and microcell systems and we will clearly mark the new features we must address with microcells. The following step is then to state the main objectives to be achieved with the cellular design and to present a review of the classic cell design techniques for uniform systems.

After this introductory material, we will focus our study on personal communications systems (PCS) by considering real cases. In particular, we will show:

1. The cellular coverage and cellular architecture in microcell and picocell scenarios;
2. The compatibility matrix from the propagation losses with and without power control;
3. The demand vector to be satisfied by each BS (base station) considering its surface and the traffic demand per surface unit;
4. Channel planning using fixed-channel assignment (FCA) algorithms, and

5. Channel planning using dynamic channel assignment (DCA) algorithms.

This chapter is not intended to give a general tutorial review of the different techniques for the channel planning design of cellular systems. The reader is encouraged to review the excellent papers available with comprehensive tutorials [3–15]. Instead, we focus on some techniques applicable for urban microcellular systems.

6.2 Microcellular Design Terminology

So as not to interrupt the presentation of the different topics to be treated in this chapter with the definitions of the terms used in cellular design, we introduce the main terms used throughout at this point:

Adjacent-channel interference: When two channels are close, they can suffer from so-called "adjacent-channel interference." There are many ways to reduce adjacent-channel interference: by improving the radio-frequency bandpass filters of the channel, by power control, by changing the modulation, etc. The interference channel affects mainly the co-users of a cell [2].

Blocking probability: This is the proportion of unattended requirements for channel assignments. We can distinguish between blocking probability for new calls and for handoffs.

Cell cluster: A cell cluster is a set of cells in which all the available channels are used and none of them are reused. Many cellular systems are designed by repeating a cell cluster along the system [2, 7]. In this case, the cell cluster is the basic element of the system architecture. Often in microcellular systems the cell architecture is quite far from being a uniform architecture based on the cluster concept, nor does the cell cluster itself make sense [4, 5, 10].

Cell coverage area: This is the actual cell area for which the field strength is greater than a certain threshold level which can refer to the receiver sensitivity noise level or to an interference power level [1].

Channel: Any method that allows the bidirectional transmission of a voice or data signal. There are many channel forms [2, 6]: a dedicated frequency band, such as in FDMA (frequency division multiple access); a dedicated slot of time for a temporal sequence defined over a single frequency carrier, such as in TDMA (time division multiple access); a code sequence using a spread-spectrum technique, such as in CDMA (code division multiple access); a sequence of frequency hops as in FHMA (frequency-hopping multiple access); a combination of some of the previous techniques, such as GSM (global system for mobile communications), where TDMA-FDMA is used [7], etc. In this chapter, we assume that the channels are sorted and that each one of them has an order number defined that adheres to the criterion that the adjacent-channel interference between a pair of channels decreases with the difference of the order numbers assigned to these channels.

Channel management charts: These are the relations of the channels assigned to each cell of a system using FCA [2, 11].

Carrier-to-interference ratio (C/I): The quality of the received signal depends on the C/I ratio. The minimum value for the C/I, which guarantees an acceptable signal quality, is a very important design parameter. This minimum varies with the modulation and with the multiple access technique employed. A value of about 10 dB is a conservative figure for digital mobile communication systems (a value of 9 dB is recommended for GSM), whereas analog systems can require about 20 dB [2, 9].

Cochannel interference: When two or more subscribers are using the same channel, they can suffer from so-called "cochannel interference." No protection against this interference is available except to use the power control or adaptive antennas or to move the subscribers further away. The minimum distance for the cells where a channel can be reused is the termed the *reuse distance* [2].

Compatibility matrix: In a cellular system, this is a squared matrix that has as many rows or columns as cells in the system. The term c_{ij} indicates the minimum threshold in the difference of the channel order numbers assigned to cells i and j, which guarantees compatibility among the users of these channels [4, 12, 13]. As we shall see later, the terms of the compatibility matrix are computed, considering a prescribed percentage of locations in the cells, a minimum value for the C/I ratio. A zero value for the term c_{ij} indicates that the same channels can be reused in cells i and j. Small values for the matrix terms make channel assignment easier. Microcell systems with high digital protection against interference can have a compatibility matrix with all of its terms null except for the diagonal terms that take the value of one. This last situation is ideal for the design system because any channel can be assigned to any cell without interference problems.

Demand vector: Each term of this vector corresponds to a cell of a mobile system and its value is the traffic in Erlangs demanded by the cell [14], which meets the prescribed GOS requirements.

Grade of service (GOS): The GOS parameter is directly related to the blocking probability. Usually, the GOS is directly given by the percentage of the blocking probability of the system. Many times the weight in the GOS definition of the blocking probability for handoffs is substantially greater than for new calls.

Handoff: When the MS (mobile station) moves away from the coverage area of a cell, it will need to be assigned a new channel from the base station (BS) of the cell into which it is going. In microcellular systems, several handoffs per call can be foreseen [1].

Power control: Using this technique in a channel when the received power is greater than a given threshold level, the transmitter reduces the total radiated power until the received power is close to this level. In this way, the interference caused by this transmitter is reduced [7, 8].

Spectrum efficiency: This parameter indicates the degree of optimization achieved using the radio spectrum in cellular systems. It is measured in Erlangs/(MHz $\times$ km^2). The spectrum efficiency depends on the number of cells, the digital technology employed, and, of course, on the quality of the cell design applied [2, 9].

Tessellation schemes: In systems with uniform cells, the design can be performed by a tessellation of cell clusters [2–11].

Traffic distributions: In a cellular system the traffic distributions are mainly characterized by the rate of channel requests to cope with new calls or handoffs [14]. The number of channels required to handle the traffic with a prescribed grade of service is given by the demand vector.

Transition probability matrix: The entry m_{ij} of this matrix represents the probability of an MS moving from cell i to j [13].

Trunking efficiency: This parameter indicates the efficiency of the shared use of the available channels by a set of MSs. This set can be formed by the channels of a BS or those of the complete systems [2, 9]. Poor trunking efficiencies are obtained when the number of channels to share is less than 20. Usually, no significant variations in the trunking efficiency appear when the number of available channels in the set is greater than 20.

6.3 Historic Evolution of Cell Planning

Historically, wireless communication systems started with the deployment of minicells, each one of them covering a relatively large area, for instance, with cell radii of several tens of kilometers. The BSs were located on the tops of hills, towers, and the highest buildings in the cities. The BSs and the MSs were capable of radiating a relatively large amount of power and were able to operate in a large dynamic range. In most cases, the power coverage could be approximated by a log-normal distribution, which means value depends on the distance d, between the BSs and the MSs following a d^{-n} power law, where the exponential coefficient n usually takes a value between 3 and 5. A traffic demand constant in each cell was also considered as a first approach for system planning. The technology employed was analog (except for the signaling data), frequency modulation (FM) was preferred, and the FDMA was used to attend to different calls simultaneously. Due to the large size of the cells, the transit between different cells of an MS was relatively scarce so the rate of channel assignments for handoffs was a small fraction of the rate of the channels assigned for new calls. Because the transition of an MS between cells was slow, a relatively large amount of time was available for the handoff process. In this situation a cellular architecture based on a hexagonal system with an FCA scheme is the best solution [2, 7].

The situation was ideal from an academic point of view: A cluster of cells could define the cellular system. The cluster repeats itself, forming a tessellation that covers the complete system. Each cell in the cluster has been assigned a fixed-frequencies set (fixed channels). This set of frequencies repeats itself in the corresponding cells in the rest of the clusters that form the tessellation, leading to the reuse channel concept.

Using the hexagonal tessellation just outlined, it is possible, in principle, to attend to an unlimited number of users if the number of clusters in the tessellation can be incremented by simply reducing the size of the cells.

Of course, in the real world [1], the actual shape of the cells is not perfectly hexagonal; the size of the cells is defined by the area where the power coverage is greater than a given threshold level. Not all the cells have the same area, nor do they have the same traffic demand. These variations from the regular plan could be corrected by small changes (tuning) in the channel plan or in the cell coverage by tilting or changing the antenna and so on [15]. Another possibility is to design cell meshes with a nonuniform cell size. The idea was to cover the more populated areas such as the centers of cities or highways with denser cellular meshes [16–18]. An overlay of local meshes of small cells covering the more populated areas of the system with a mesh of large cells, which covered the complete territory, was deployed. Still, the cells were large enough to allow us to compute propagation losses using the exponential law or some other empirical approach where the losses depended only on the distance between the MS and the BS. Again, the average number of handoffs per call is low and the academic way of designing the system can be applied, although with a little more effort in the final tuning process.

However, when the traffic demand increased, especially in the downtown area of large cities, in large public buildings, etc., very small cells had to be implemented; actually we can speak of micro- or even picocells, with sizes ranging from some tens to a few hundred meters. In these cells, the empirical laws for computing the propagation losses became nonapplicable and due to the small size of the cells, several handoffs per call can be foreseen. This implies that some new concerns must be undertaken:

- A detailed design for the antenna locations, for antenna pointing, and for the transmitted power by BSs should be performed to provide appropriate power coverage.
- A detailed computation or estimation of the compatibility matrix of traffic distribution and of the transition probability matrix should be obtained for the channel planning algorithms and system simulation [4].
- Tailored channel planning should be implemented if an FCA [19, 20] is chosen. Now the system cannot be obtained as a tessellation of a cellular cluster that repeats itself [21]. Many times a DCA [22, 23] is preferable to an FCA because the DCA fits better for the short-term fluctuation of traffic demand. At other times, the hybrid channel assignment (HCA) is selected because it behaves better in systems with a high degree of mobility where the blocking probability for handoffs must be minimized [24, 25].

6.4 Cell Design Objectives

The objectives of a cellular design are very easy to state:

1. The electrical coverage of the cell area must be guaranteed. The received field must always be above a threshold level and holes in the coverage should be avoided. Coverages with a constant field value in all the cellular area are also desirable. Using the power control technique [8] in both the BS and the MS, uniform field coverage can be achieved in almost all of the cell area. Limitations in the dynamic range of the power control give coverages that are uniform only in a part of the cell. Figure 6.1 shows the typical power coverage for a cell situation where the path losses are only a function of the distance between the BS and the MS.

   The power coverage should be above the receiver sensitivity threshold in a maximum possible percentage of points of the cell, for

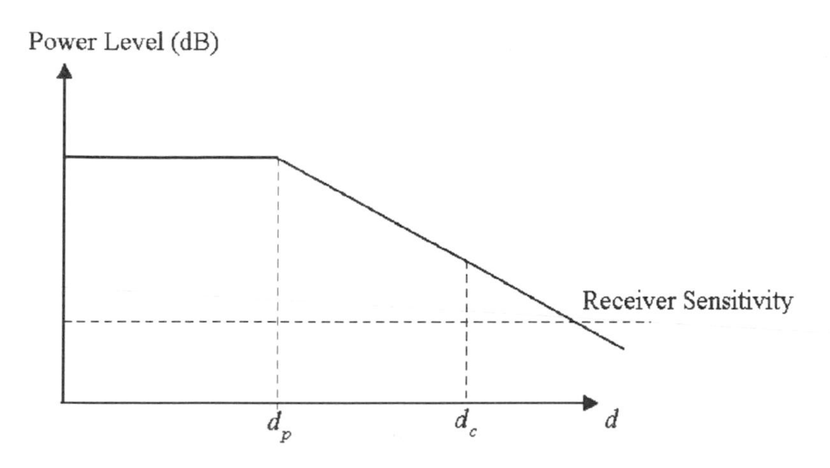

Figure 6.1 Shape of the signal strength as a function of the distance between the BS and the MS assuming that the path losses depend only on this distance and that a power control technique is used. The signal strength is constant for $d < d_p$, just in the range where the power control works. In all the cell area, $d < d_c$, the signal strength is above the sensitivity level of the receiver.

example, greater than 90%. So as not to interfere with other cells, the field strength of the signal of each BS outside its cell must be reduced.

2. The C/I ratio in a large percentage of the cell area (e.g., 90% or 95% of the points in the cell) must satisfy a minimum of quality in the received signal. The compatibility matrix will give the restrictions for the channel that will guarantee the minimum threshold of signal quality.
3. We want to reach a GOS that provides, in each cell, enough channels to satisfy its traffic demand for new calls and for handoffs. Usually, the blocking probability for handoff assignments should be much less than the blocking for new calls.
4. We must minimize the number of BSs so as to reduce the cost and the time of cellular deployment. The design approach must be flexible in order to locate the BS transmitters (antennas) at the available sites. The number of available sites is quite limited and in many cases is strongly defined by nontechnical reasons.
5. The radio system should be integrated with the land-fixed network. This usually means that the microcellular design should fit some requirements to connect correctly to a previously deployed land-fixed network.

6.5 Design of the Geometry of the Cellular System

Traditionally, the hexagonal shape has been chosen for cellular geometry in the design of systems with macrocells [2, 11]. In these cases, for which the propagation losses can be approximated by the exponential law and where the traffic can be assumed uniform, the hexagonal cellular system is the optimal solution, providing that the BSs are located on the nodes of the hexagonal mesh. Depending on the values of the exponential coefficient that governs the path losses and on the allowed C/I ratio, different cell clusters, tessellation schemes and channel management charts can be obtained for the design of cellular systems. There are many references where the reader can find excellent presentations of all the steps for the design of cellular systems with hexagonal meshes [2, 7, 11].

When we are dealing with microcellular systems, things are quite different, because the exponential law does not apply anymore. One might choose to use, as a preliminary design, the hexagonal approach and then, after successive modifications and tuning steps, obtain the microcellular geometry. However, we can foresee that the final cellular geometry will be quite far from a uniform hexagonal mesh. In addition, we must remember that there are many nontechnical restrictions for choosing the antennas sites. So, probably the best way is to estimate the areas of the cells and just start to locate the antennas in the allowable sites following some simple design rules. The cell areas can be estimated considering the foreseeable traffic per surface unit and the desirable number of channels per cell that guarantees a minimum of trunking efficiency. After this first BS location, an analysis of the performance of this preliminary cell design should be performed. Should this design not accomplish the desired specifications, a new step in the cell design should be performed and the process must be repeated until the design becomes acceptable. In the steps of this iterative process, one can vary the location of the antenna sites, the antenna radiation patterns, the antenna orientation, and the power radiated by each BS. This is a complex process that in the future could probably be performed by advanced optimization algorithms such as the genetic ones running in computers with high-paralleled architectures.

Nevertheless, the extreme difficulty of the problem when we are dealing with an irregular downtown area with a large variety of street widths, heights of skyscrapers, and sizes of open areas, requires guidelines for the microcellular designer. We start by assuming an example of a uniform urban area, like the one indicated in the 3D view presented in Figure 6.2.

We can consider the urban environment of Figure 6.2 to be representative of a uniform area. Lots of measurements and computations of propagation losses for wireless applications have been performed for this area. They are

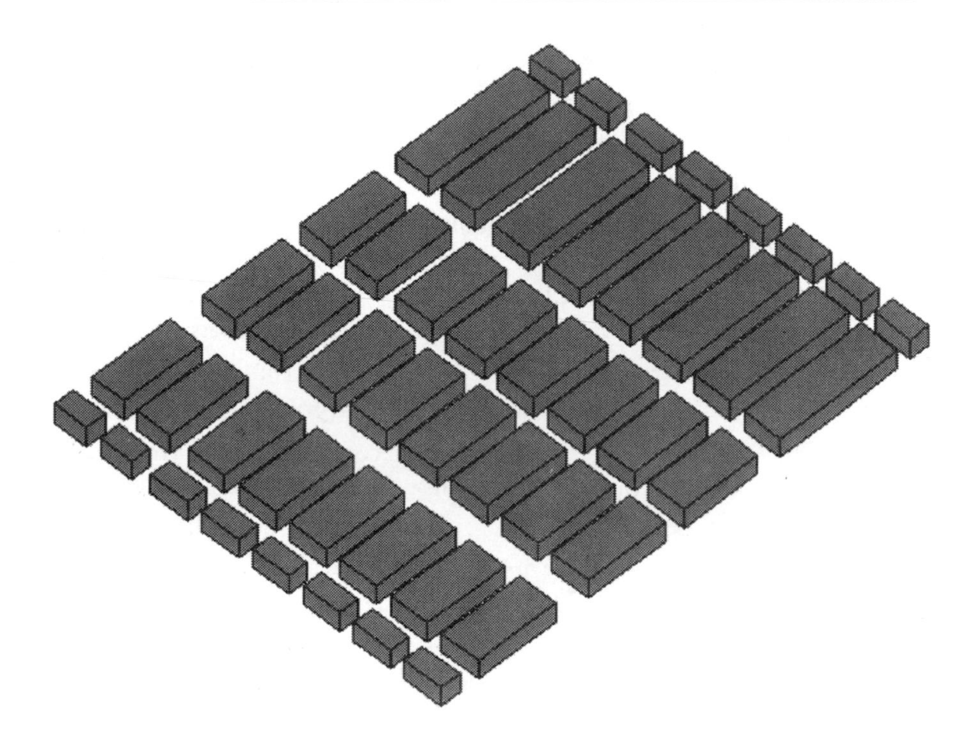

Figure 6.2 Three-dimensional view of part of the Manhattan area indicated in Figure 2.13. The surface area of this section is 740 × 600m, the street width varies between 20m and 30m, and the height of the buildings has been approximated by a uniform value of 25m. With this approach, this scenario can be considered as an example of a uniformly distributed urban area.

well documented in the literature [26, 27]. We have considered several possibilities for the BS sites, and we have analyzed the power coverage for each one of these sites using Fast Propagation Analysis Computer Tool (FASPRO) [28, 29], a well-validated computer tool for the analysis of propagation in urban scenarios. FASPRO uses a completely three-dimensional GTD/UTD [30, 31] propagation model. Because FASPRO considers the diffraction on all the edges and the coupling mechanisms up to a third order (e.g., reflection-diffraction-reflection), its results are quite reliable. FASPRO can compute the power of each polarization component of the field (x, y, and z components). Details about the validation of FASPRO can be found in [29]. In all of the cases presented here, the computations have been obtained using these parameters:

- The BS is radiating a total of 1 mW (0 dBm).
- The height of the BS antenna is 4.5m.

- A short antenna of 2 dBi of gain is used for the receiver, 1.5m high. The sum of the power of each one of the three Cartesian polarization components has been considered.
- A frequency of 922 MHz is used.
- The received field level is obtained considering all of the simple and double-order GTD/UTD mechanisms (e.g., diffractions, double-reflections, reflection-diffractions, etc.) and the third-order mechanism when one of the effects in the triad is a ground reflection.
- The field has been computed in a mesh of 100 × 100 observation points, uniformly distributed in the urban scene of Figure 6.2.
- At each observation point, the field is computed as the average of the GTD/UTD predicted values at this point and at the four closest points.

Following a scheme for the BS site location close to that of [32–34], we have considered four cases. In the first case the BS is just in the center of a cross street. The results of the power coverage for the cross street case with the omnidirectional antenna are shown in Figure 6.3 using a gray code. We can see the symmetry and homogeneity of the coverage and the fact that a field level greater than −130 dBm (which corresponds to a propagation loss

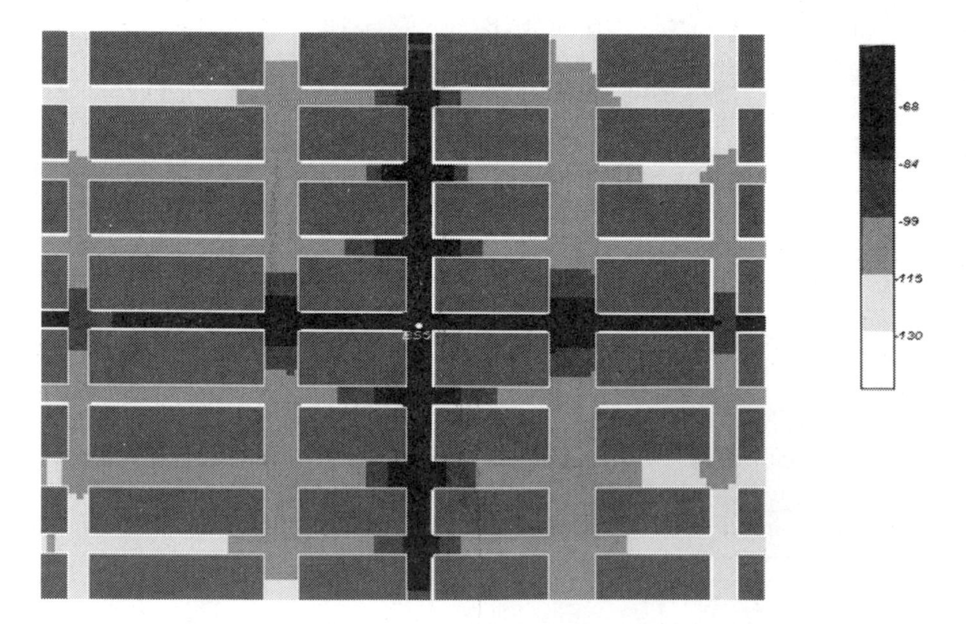

Figure 6.3 Power coverage (dBm) of one BS for the cross street case with omnidirectional antenna.

of −130 dB, because the BS is transmitting 0 dBm) extends more than two street blocks away from the BS in the horizontal direction and three cross streets in the vertical direction. The area with the stronger level, say, greater than −100 dBm, extends up to one block in the horizontal direction and one and a half cross streets for the vertical one. In general, the field level is large in the streets in the line of sight (LOS); however, the power coverage is quite low in all the streets that are in the non-line of sight (NLOS) in the transmitter, even in the closest ones. Similar behavior for transmission losses was described in [33] for this same scenario. As reported in [33], the contours of constant losses are concave, diamond-shaped lines, centered on the antenna location and with their main axes in the two perpendicular streets that define the cross street. A parametric model for the losses was found in [33].

Figure 6.4 shows all of the cross streets of the urban scene we are considering numbered. The simple microcellular architecture that locates a BS with an omnidirectional antenna in each cross street will guarantee very good power coverage; however, it will give a very bad compatibility matrix. (Most of the terms of the matrix take high values; see Appendix 6C for the computation of the interference matrix.) Of course, it is not necessary to cover all the cross

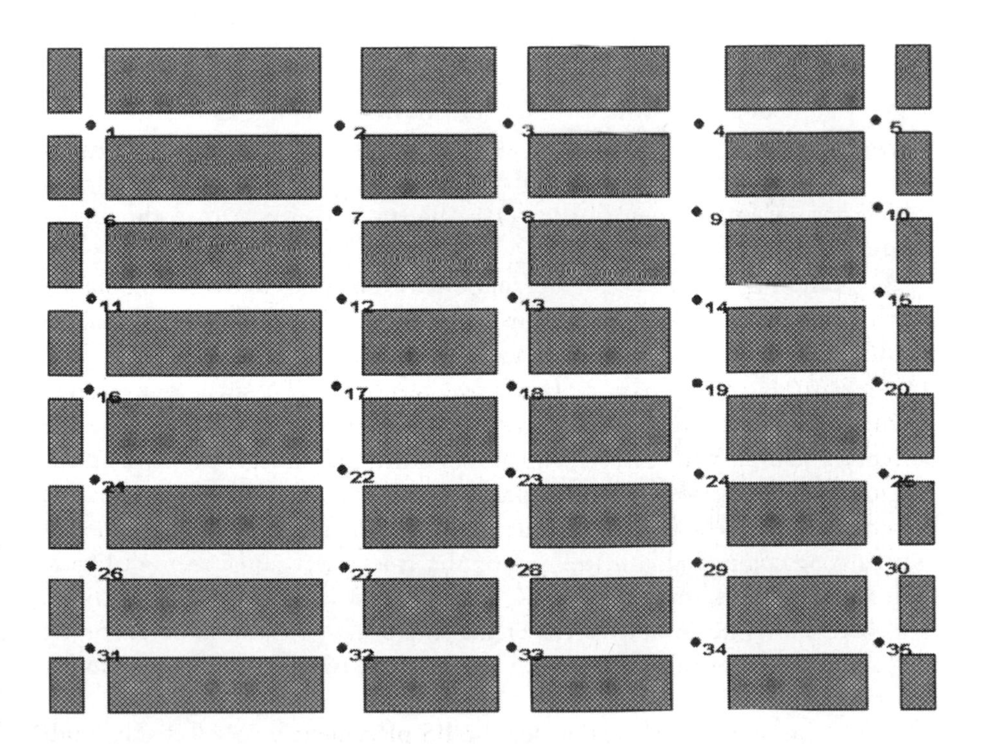

Figure 6.4 Cross street antenna site numeration.

streets with one BS. We have many possible schemes for covering only some of the cross streets.

A less dense scheme is to locate one BS in the cross streets with numbers: 1, 3, 5, 16, 18, 20, 31, 33, 35 (scheme CO1). This scheme gives a compatibility matrix that is not too bad, (several terms are zeros and the others are not much higher), although in the NLOS areas (e.g., between cross streets 13 and 15), a low coverage level can be foreseen as we can see in Figure 6.5(a), where the coverage levels are represented assuming that all the BSs of scheme CO1 are active. The area covered by each BS is pictured in Figure 6.5(b) considering the best server rule: Each observation sampling point is associated with the BS that provides the highest level in the power coverage.

Probably, one of the best schemes using omnidirectional antennas can be to locate the BSs in alternate cross streets, for instance, in the intersections: 1, 3, 5, 7, 9, 11, 13, 15, 17 19, 21, 23, 25, 27, 29, 31, 33, 35 (scheme CO2) or the less dense scheme: 1, 5, 8, 11, 15, 18, 21, 25, 28, 31, 35 (scheme CO3). Using schemes CO2 and CO3, any point in any street is in the LOS and close to one BS, and, quite infrequently, the point is radiated by two or more BSs. This will mean favorable compatibility matrices and good power coverage in all of the urban areas considered as shown in Figures 6.6 and 6.7.

A way of reducing the interference between adjacent BSs and therefore improving the quality of the compatibility matrix is to use directive antennas. Figure 6.8 shows the power coverage of one BS radiating as in the case of Figure 6.3, except that now the antenna is directive and it is pointing toward the right. Figure 6.9 shows the radiation pattern of the antenna considered, which has a gain of approximately 10 dBi and a 3-dB beam width of 33 degrees.

Examining Figure 6.8 we note that the high field strengths extend mainly in the antenna pointing direction, while in the rear area of the antenna, the field decreases very fast. The radiation level in the NLOS area is quite low. In a few words, this BS has very good coverage in the LOS street that emerges from the BS cross street: up to three or four blocks in the street where the antenna's main axis is located, and up to about one cross street in the perpendicular street (vertical ones). The fact that this antenna concentrates its radiation in a moderately well-defined zone can be used to define microcellular schemes with a good compatibility matrix. Using the cross street numeration of Figure 6.4, we can suggest the schemes CO1, CO2, and CO3. Figures 6.10, 6.11, and 6.12 show the coverage areas of the microcells of these schemes. As we will see in the following section, these schemes also provide good compatibility matrices.

Of course, another solution for the BS placement is to select the "midpoints" between cross streets. Figures 6.13 and 6.14 show the power coverage

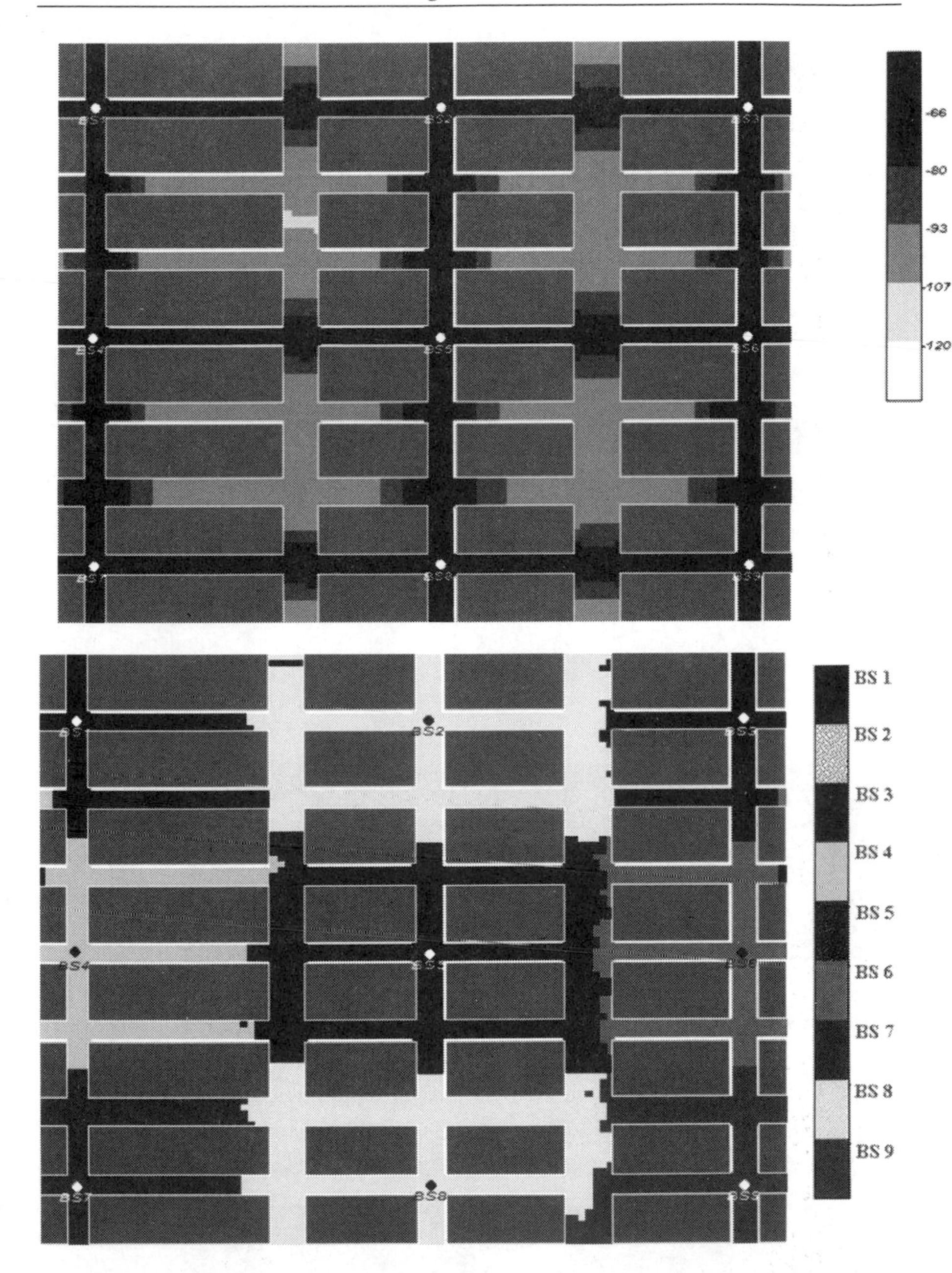

Figure 6.5 (a) Power coverage levels (dBm) for the BS scheme C01 considering the power received from the best server only at each point; (b) area covered by the cells of the C01 scheme using omnidirectional antennas and considering the best server at each point.

Figure 6.6 Area covered by the cells of the CO2 scheme using omnidirectional antennas and considering the best server at each point.

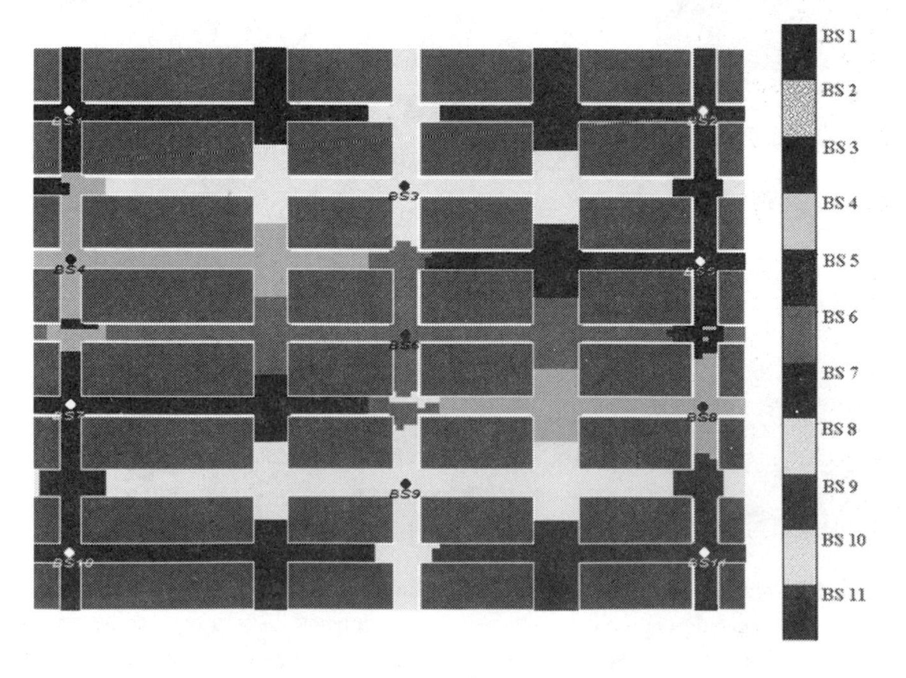

Figure 6.7 Area covered by the cells of the CO3 scheme using omnidirectional antennas and considering the best server at each point.

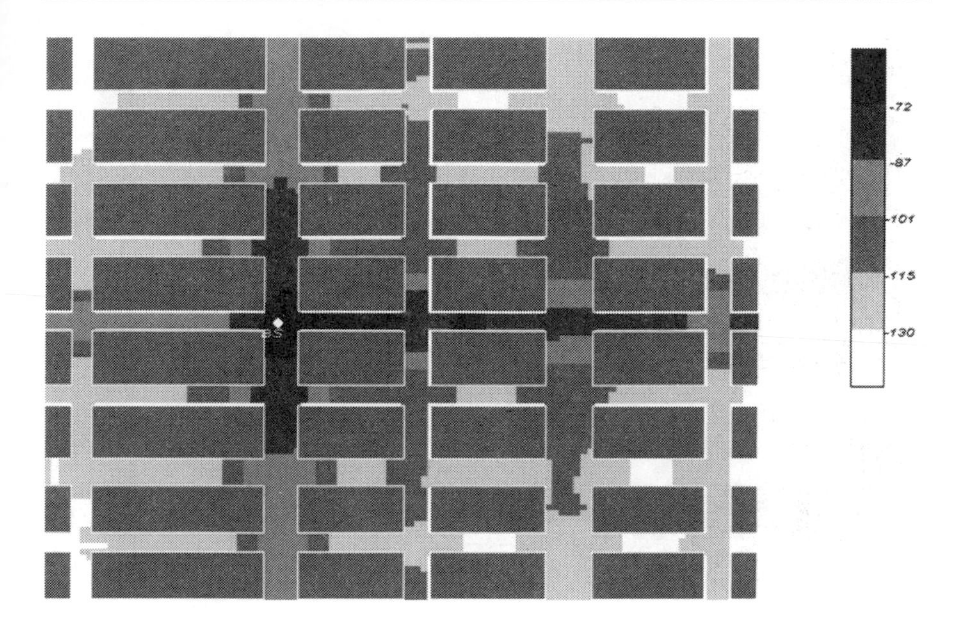

Figure 6.8 Power coverage (dBm) of one BS for the cross street case with directive antenna.

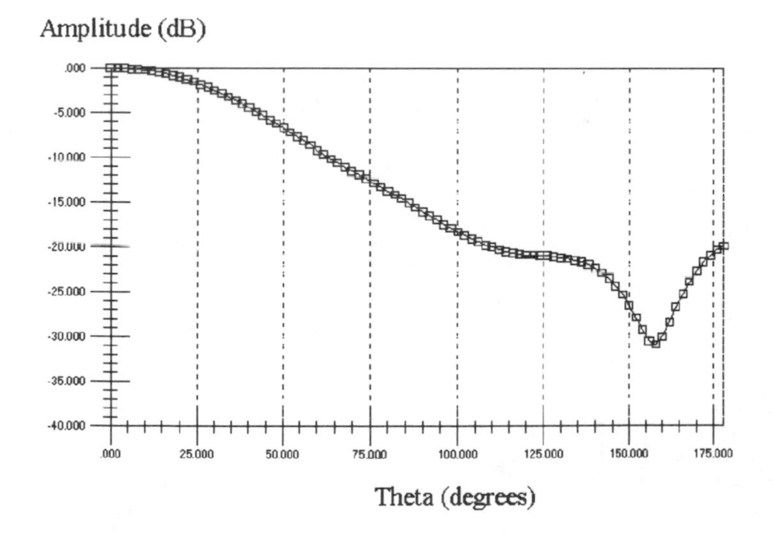

Figure 6.9 Radiation pattern of the directive antenna used in the coverage analyses.

using omni and directive antennas, respectively. A general comment for both figures is that the radiated energy is concentrated in the BS street, especially in the directive antenna case. Little energy goes to the perpendicular street. With these BS locations, it is difficult to find microcellular schemes that

Figure 6.10 Area covered by the cells of the CO1 scheme using directive antennas and considering the best server at each point.

Figure 6.11 Area covered by the cells of the CO2 scheme using directive antennas and considering the best server at each point.

Figure 6.12 Area covered by the cells of the CO3 scheme using directive antennas and considering the best server at each point.

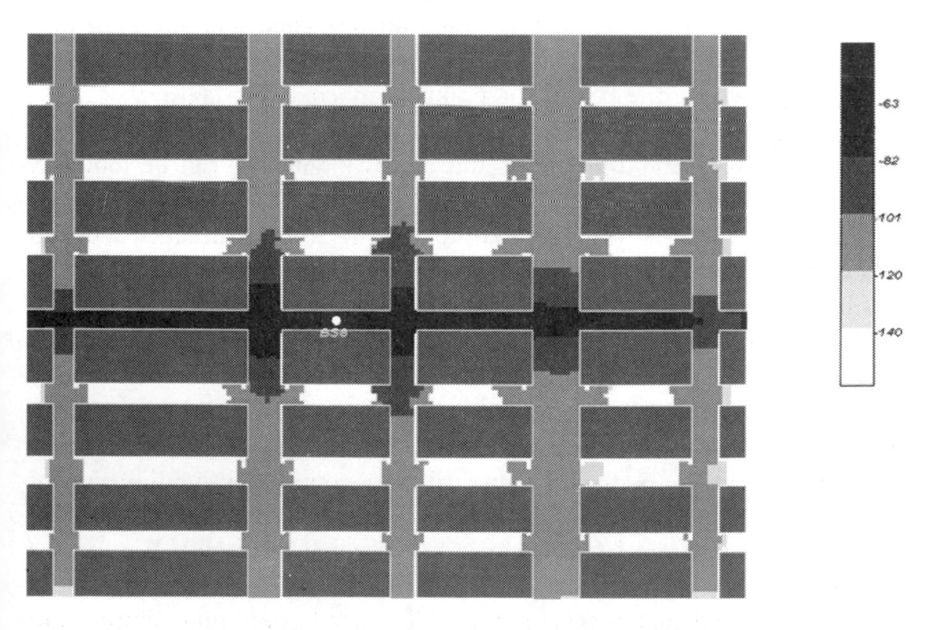

Figure 6.13 Power coverage (dBm) of one BS for the midstreet case with omnidirectional antenna.

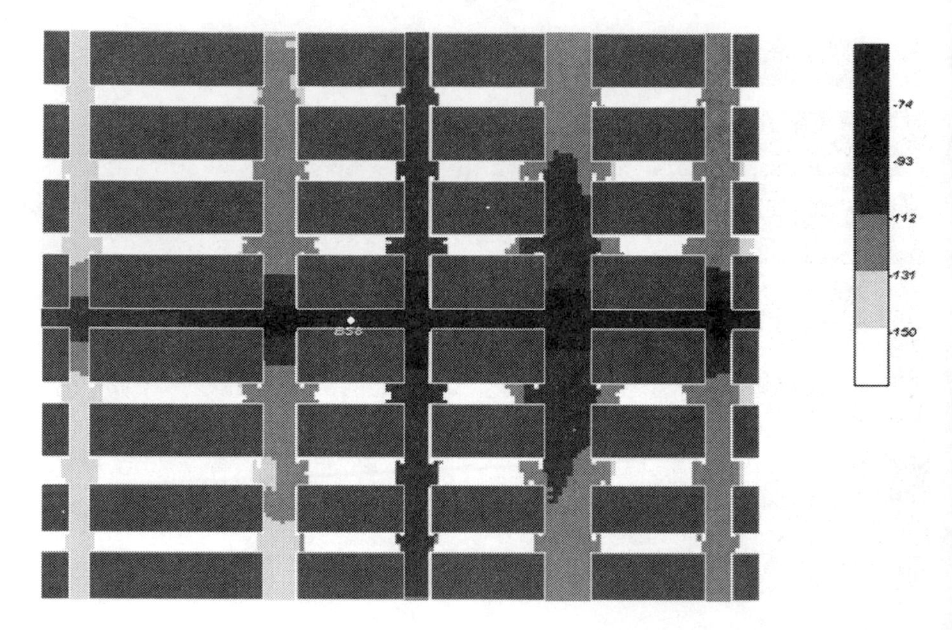

Figure 6.14 Power coverage (dBm) of one BS for the midstreet case with directive antenna.

simultaneously provide good power coverage and an acceptable compatibility matrix. However, with some effort we can find very good solutions. The new scheme is defined considering the cell site numeration of Figure 6.15, in particular by locating the BSs in the points numbered: 1, 3, 6, 8, 9, 11, 14, 16, 17, 19, 22, 24, 25, 27 (M1 scheme). A picture of the microcellular area coverage of this scheme using omni and directive antennas can be found in Figures 6.16 and 6.17, respectively.

Until now we have analyzed and proposed several microcellular schemes for a quite uniform urban scenario in Manhattan, shown in Figure 6.2. But what happens when we have an irregular urban scenario, as can be, for instance, the case for the center of Madrid, pictured in Figure 2.17. For this new case, we can try to use the best schemes obtained for the uniform case. Figure 6.18 shows, in a horizontal cut of the scenario, the BS sites chosen to obtain a combination of the microcellular schemes, CO3 and M1. We will call the new BS scheme I1. Figure 6.19 shows the cell areas considering omnidirectional antennas. The same cells are presented in Figure 6.20 considering directive antennas (pointing from left to right in a horizontal direction). We can see that the schemes are, as first solutions, quite good. Now by further tuning the cell sites or the radiated power for each BS, a nearly optimum solution can be found for the power coverage.

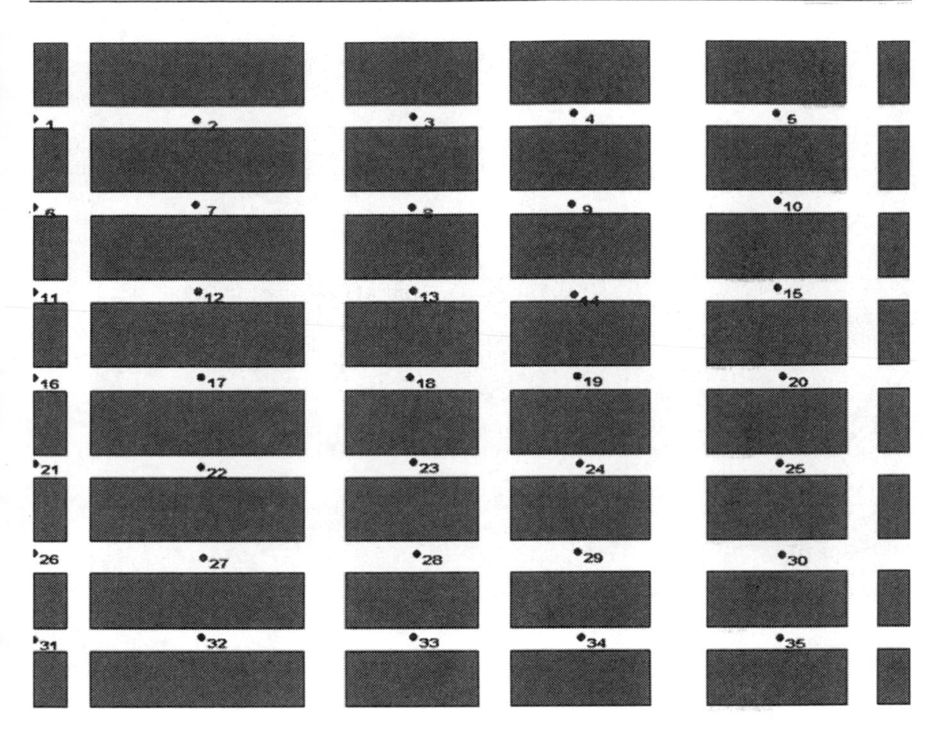

Figure 6.15 Midstreet antenna site numeration.

6.6 Analysis of the Compatibility Matrices for Some Microcellular Systems

In this section, we present some results and conclusions of an interference analysis of the BS localization schemes described in the previous section. In particular, values for the cell areas, interference matrices, and compatibility matrices are shown in Tables 6.1 and 6.2 for the cases CO1 and I1, both with omnidirectional antennas. These values have been obtained using the expressions of Appendixes 6A, 6B, and 6C for the computation of the microcell areas, the interference, and the compatibility matrices terms, respectively. The expressions have been applied considering these data: (1) a value of 18 dB for the threshold level of the C/I, which guarantees compatibility; (2) 7.5 dB per octave for the slope of the front-end filter, and (3) the propagation losses generated by code FASPRO [29], described in the previous section. If we examine the aspects of the matrices of both tables, we can conclude that it is difficult to observe any uniform behavior in the form of the matrices and it is not possible to deduce a relation describing how the different terms of the matrices vary. This is especially true in the case of Table 6.2, which presents

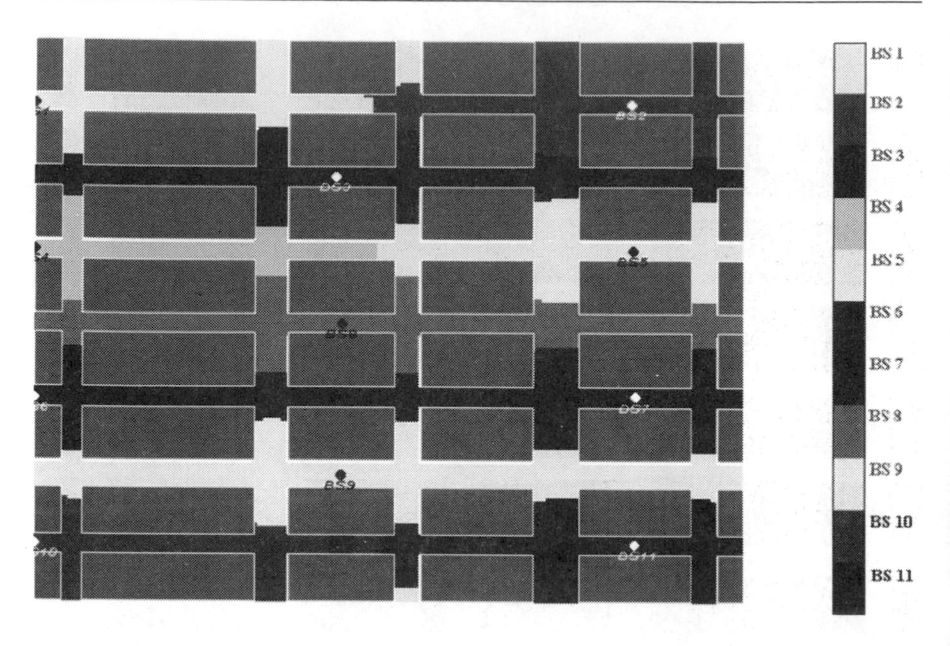

Figure 6.16 Area covered by the cells of the M1 scheme using omnidirectional antennas and considering the best server at each point.

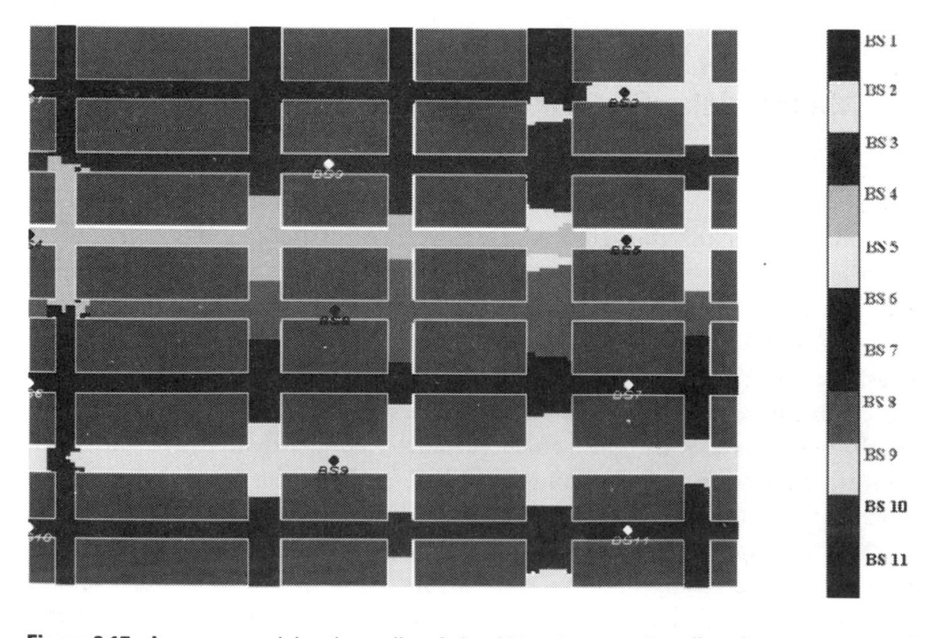

Figure 6.17 Area covered by the cells of the M1 scheme using directive antennas and considering the best server at each point.

Figure 6.18 Cross street antenna site location for the irregular urban case, I1.

results for a quite irregular urban scenario. More of the asymmetries of Table 6.1 are due to the small asymmetries in the urban geometry. In fact, the antennas are not located exactly in the centers of the cross streets for all the BS locations indicated in Figure 6.5. In a real-world application, one must expect these kinds of asymmetries. Summarizing, we can state that in microcellular systems, the compatibility matrix is usually quite irregular and therefore the channel assignment algorithms should be able to manage these kinds of arbitrary compatibility matrices.

We can use the compatibility figures in order to measure the quality of the compatibility matrix of a BS localization scheme. We define a compatibility figure as the adding up of a set of elements of the compatibility matrix. The units of the compatibility figures are expressed in *channels of separation,* the same units used for the compatibility matrices. If we consider one BS as a reference in any one of the uniform BS schemes of Figure 6.5, we can define three compatibility figures:

1. The compatibility figure for the first tier of BS. This tier being defined by the four closer BSs to the reference BS, and located in the endpoints of a Latin cross (+), which has its center in the reference BS. The

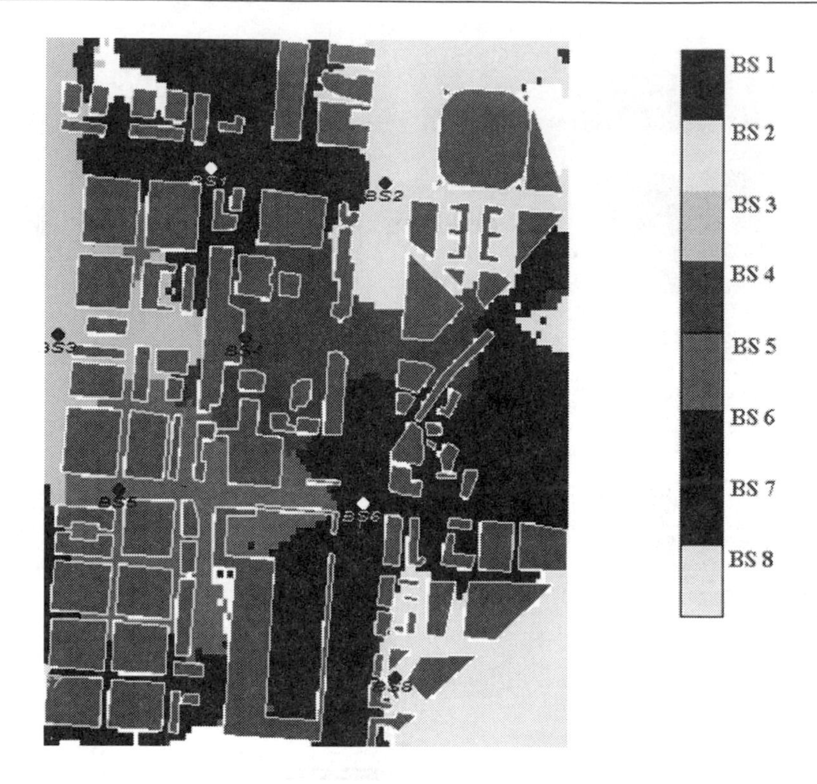

Figure 6.19 Area covered by the cells of the I1 scheme using omnidirectional antennas and considering the best server at each point.

compatibility figure is the adding of the four terms of the compatibility matrix that relate the BSs of the tier with the reference BS.

2. The compatibility figure for the second tier of BSs. This figure is computed in a similar way to that of the first tier, except now the four BSs of the tier are located in the endpoints of a Greek cross (×).
3. The total compatibility figure, defined as the adding of the figures for the first and second tiers.

To obtain representative values, the compatibility figures have been averaged considering several BSs as reference. Table 6.3 shows the compatibility figures for the BS configurations described in Section 6.5. In the I1 scheme of Table 6.3, corresponding to the irregular urban scenario, it is difficult to distinguish the BS of the first or second tier. For these last cases, the total compatibility figures have been computed considering as a BS reference, the one that has the highest figure value.

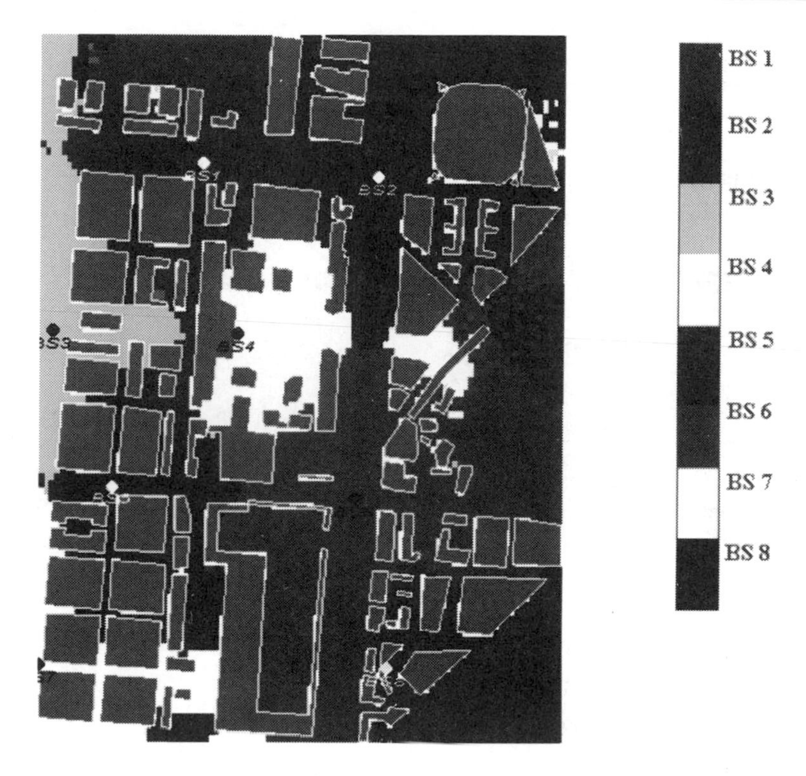

Figure 6.20 Area covered by the cells of the I1 scheme using directive antennas and considering the best server at each point.

Analyzing the compatibility figures of Table 6.3 we can conclude that, in general, they are quite acceptable, especially for the cases M1 and I1. These surprisingly low compatibility figures have been obtained thanks to our experience of the power coverage of omnidirectional and directive antennas in different locations, as discussed in Section 6.5

6.7 Fixed-Channel Assignment in Microcellular Systems

As we have stated in previous sections, an FCA is usually a complex problem in microcellular systems due to the high degree of inhomogeneity in the traffic parameters and in the compatibility matrices of these systems. The problem, also known as the channel assignment problem (CAP), is quite difficult because the allowable spectrum is very limited in wireless communications. A CAP can be formulated as follows: Given a system with N_T microcells, the goal is to obtain channel management charts that, using a minimum of channels, satisfy

Table 6.1
Detailed Results of a Compatibility Analysis for Case C01 With Omnidirectional Antennas (Analyzed in Figure 6.5)

NUMBER OF CELLS: 9
CELLS AREA (km^2):

.01169	.03054	.01059	.01651	.03898	.01623	.01426	.03201	.01591

INTERFERENCE MATRIX (dB)

.000	2.639	14.80	4.077	11.82	18.62	10.80	15.50	22.61
3.656	.000	1.921	7.210	4.433	10.81	16.23	12.37	16.74
11.78	1.234	.000	16.73	8.693	5.963	22.24	16.82	14.07
4.844	11.51	24.84	.000	1.837	12.20	4.560	10.76	21.69
7.875	2.84	15.86	3.362	.000	.775	6.771	3.977	10.36
14.80	9.60	4.50	11.80	1.004	.000	14.24	10.29	4.970
11.84	15.88	24.01	5.568	9.345	15.57	.000	2.233	15.22
16.69	11.75	18.17	9.648	3.988	10.98	3.532	.000	.589
23.26	14.59	15.78	16.47	11.114	4.644	12.16	.963	.000

MATRIX C [channel spacing, 90% of cell Ci/Ii < 18.0 dB] :

2.6	2.1	1.0	1.8	1.3	1.0	1.0	1.0	.0
2.1	2.6	2.4	1.4	2.0	1.1	1.0	1.0	1.0
1.0	2.4	2.6	1.0	1.2	1.7	.0	1.0	1.0
1.8	1.4	1.0	2.6	2.2	1.0	1.7	1.1	1.0
1.3	2.0	1.2	2.2	2.6	2.5	1.4	1.8	1.0
1.0	1.1	1.7	1.0	2.5	2.6	1.0	1.0	1.7
1.0	1.0	.0	1.7	1.4	1.0	2.6	2.1	1.0
1.0	1.0	1.0	1.1	1.8	1.0	2.1	2.6	2.5
.0	1.0	1.0	1.0	1.0	1.7	1.0	2.5	2.6

the demand vector, $V(M)$ and the compatibility matrix, $CM(M, N)$ of this system.

A CAP is a nonpolynomial (NP) complete problem. That means that due to the difficulty of the problem, it is not possible to implement an algorithm that solves the problem with a CPU time that can be written in terms of a polynomial of N_T. On the contrary, the CPU time will increase exponentially with N_T.

Due to its complexity, most of the algorithms find a *heuristic* solution for the CAP. These heuristic algorithms can be iterative [12, 35–37] or noniterative [37, 38]. Probably, the most efficient approaches for solving NP problems

Table 6.2
Detailed Results of a Compatibility Analysis for Case I1 With Omnidirectional Antennas (Analyzed in Figure 6.19)

NUMBER OF CELLS: 8 **CELL AREA (km^2):**							
.14516	.14253	.08732	.10344	.10515	.26892	.04218	.09507
INTERFERENCE MATRIX (dB):							
.000	6.771	9.941	26.085	9.935	25.533	18.889	38.600
9.620	.000	26.847	9.052	26.908	9.134	21.460	38.351
23.362	64.829	.000	55.319	15.340	70.855	5.372	81.939
21.719	9.085	14.589	.000	21.450	8.955	70.019	52.750
26.383	32.518	17.095	23.720	.000	8.049	10.932	39.739
32.949	8.716	45.953	5.214	6.294	.000	25.577	18.245
36.025	72.523	11.817	92.247	15.239	81.806	.000	81.939
71.727	57.123	86.266	51.696	83.659	16.785	72.981	.000
MATRIX C [channel spacing, 90% of cells Ci/Ii < 18.0 dB] :							
2.6	1.4	1.1	.0	1.1	.0	.0	.0
1.4	2.6	.0	1.1	.0	1.2	.0	.0
1.1	.0	2.6	1.0	1.0	.0	1.6	.0
.0	1.1	1.0	2.6	.0	1.6	.0	.0
1.1	.0	1.0	.0	2.6	1.5	1.0	.0
.0	1.2	.0	1.6	1.5	2.6	.0	1.0
.0	.0	1.6	.0	1.0	.0	2.6	.0
.0	.0	.0	.0	.0	1.0	.0	2.6

are based on the local-search techniques. These approaches are heuristic and iterative. Here, we summarize the basis of the local-search techniques. These techniques can provide a good solution for the CAP and, in addition, they are easy to implement.

The channel assignment is made assuming the worst case for the system: "All the possible calls are in progress." In these circumstances the number of active users in microcell i is given by the value of the term $V(i)$ of the demand vector. The total number of calls (active users) in the system will be

$$U_T = \sum_{i=1}^{N_T} V(i) \tag{6.1}$$

Table 6.3
Compatibility Figures for the BS Localization Schemes of Section 6.5

Microcell Configurations	First Tier	Interference Figure Second Tier	Total
C1 OMNI	8.5	4.9	13.4
C2 OMNI	7.1	4.6	11.7
C3 OMNI	5.0	4.0	9.0
C1 DIRECTIVE	6.8	2.0	8.8
C2 DIRECTIVE	4.1	5.5	9.6
C3 DIRECTIVE	5.7	5.1	10.8
M1 OMNI	3.4	4.0	7.4
M1 DIRECTIVE	4.2	4.6	8.8
I1 OMNI			5.3
I1 DIRECTIVE			7.4

The active user number j of cell i will be referred to as a_{ij}. The channels available for the system are assumed sorted and p will represent the pth channel of the sorted channel list. Usually, in FDMA systems the sorting follows an increasing frequency order.

One of the main concepts of the local-search methods is the ordered list of calls, **X**. One realization of **X** is an orderly arrangement of the users. For instance

$$X = (a_{11}, a_{12}, a_{31}, a_{22}, \ldots) \tag{6.2}$$

is a possible realization for **X**.

A list **X** shall include all the calls, therefore the list must contain U_T elements. The list **X** can be represented mathematically by the matrix X_L of dimensions $(U_T, 2)$. If call a_{ij} is in the l position of the list, we have

$$X_L(l,1) = i \tag{6.3a}$$

$$X_L(l,2) = j \tag{6.3b}$$

In each iteration of the local-search technique, the ordered list is employed to assign the channels provisionally. Usually the channel-exhaustive strategy is considered for this assignment. The exhaustive strategy starts assigning channel number 1 to the first element of the list **X**. Afterwards, one channel is assigned to the second call in **X** in such a way that the number of this channel is the smallest that the compatibility matrix, $CM(M, N)$, allows. Then, another channel is assigned to the third element of the list, looking for the channel of smaller order that allows compatibility, in accordance with the matrix $CM(M, N)$, with the previous two channels just assigned. The procedure is applied in a similar way with the fourth and consecutive calls in **X**, until all the calls of **X** have been assigned a channel number. Doing that, a compatible set of channels is assigned to the list **X**. To account for the channels assigned to the users of the list **X**, we define the matrix $F(l)$

$$F(l) = p \tag{6.4}$$

where it is assumed that channel number p is assigned to the call in the l position of the ordered list **X**. The maximum number of channels N_r (bandwidth) required for the channel assignment just performed to the list **X** is given by

$$N_r = \max[F(l)] \tag{6.5}$$

The following step in the application of the local-search technique is to check in the neighborhood $N(\mathbf{X})$ of the ordered list **X**, for a channel assignment with a lower value of N_r. A neighborhood of **X** is a set of the ordered list **X′**, such that **X** and **X′** have the calls ordered in the same way except for a small portion of calls, usually two calls, which have swapped their orders. For instance, ordered lists **X** and **X′** are neighboring if they have interchanged the values of rows l and m in their corresponding matrices of order, X_L and X_L'

$$X_L(i,j) = \begin{cases} X_L'(i,j) & i \neq l,\ m \\ X_L'(m,j) & i = l \\ X_L'(l,j) & i = m \end{cases} \tag{6.6}$$

If a list **X′** in the neighborhood of **X** is found with a lower value of the maximum number of channels required, then **X′** is chosen for the provisional channel assignment in a new iteration of the local-search algorithm. The final solution is obtained when a prefixed value for the maximum number of channels N_r is achieved or when the number of iterations in the algorithm application reaches some given limit. The different local-search algorithms differ in the way the initial ordered list is chosen, in the way the neighborhood of ordered lists is checked and in the way the algorithm is finished. Some algorithms in some iteration stages select the provisional channel assignment for the new iteration considering the first list **X′**, which has the same value for the maximum number of channels as **X**. For a more detailed discussion of the local-search algorithms, see [12, 39, 40]. The primary steps of the local-search algorithms are summarized in the flowchart shown in Figure 6.21.

To assess the ideas presented, let's consider the example of a system with three microcells with a compatibility matrix given by

$$CM(M, N) = \begin{bmatrix} 4 & 3 & 1 \\ 3 & 4 & 2 \\ 1 & 2 & 4 \end{bmatrix} \tag{6.7}$$

and one demand vector defined by $V(i) = 3, 2, 1$; for $i = 1, 2, 3$, respectively. A possible choice for one initial ordered list is as follows:

$$\text{X} = (a_{31}, a_{21}, a_{11}, a_{12}, a_{22}, a_{32}, a_{13}, a_{33}) \tag{6.8}$$

Following the exhaustive strategy to assign channels to the list **X** and taking into account the interference matrix of (6.7), we can obtain the frequency chart given by the vector $F(l)$ of Table 6.4. The bandwidth of this provisional assignment is 14, the maximum of $F(l)$. Now we check in the neighborhood of **X** for a better solution. For instance, we can consider the neighboring

$$\text{X}' = (a_{11}, a_{21}, a_{31}, a_{12}, a_{22}, a_{32}, a_{13}, a_{33}) \tag{6.9}$$

and now we make a new channel allocation following the exhaustive strategy. Table 6.4 also presents the new channel assignment, $F'(l)$. Because the bandwidth is now 13, smaller than that from the previous list, the new list **X′** should be considered for the following step in the local-search iterative process.

6.8 Dynamic-Channel Assignment in Microcellular Systems

FCA algorithms present the best system performance assuming a number of active users in each microcell equal to the value that the demand vector takes for this microcell. In this case the FCA is optimum because it maximizes the traffic in the system. However, for a particular time, the number of active users in the microcells of a system is seldom given by a fixed demand vector. Actually, the traffic fluctuates in time and from microcell to microcell; therefore, one can expect to find large variations between the number of channels required and the available channels in the FCA scheme. Large inefficiencies in the channels' distribution can appear in the FCA.

DCA was born out of the need to cope with the spatial and temporal variations in the traffic demand. Now, the idea is to share all the available

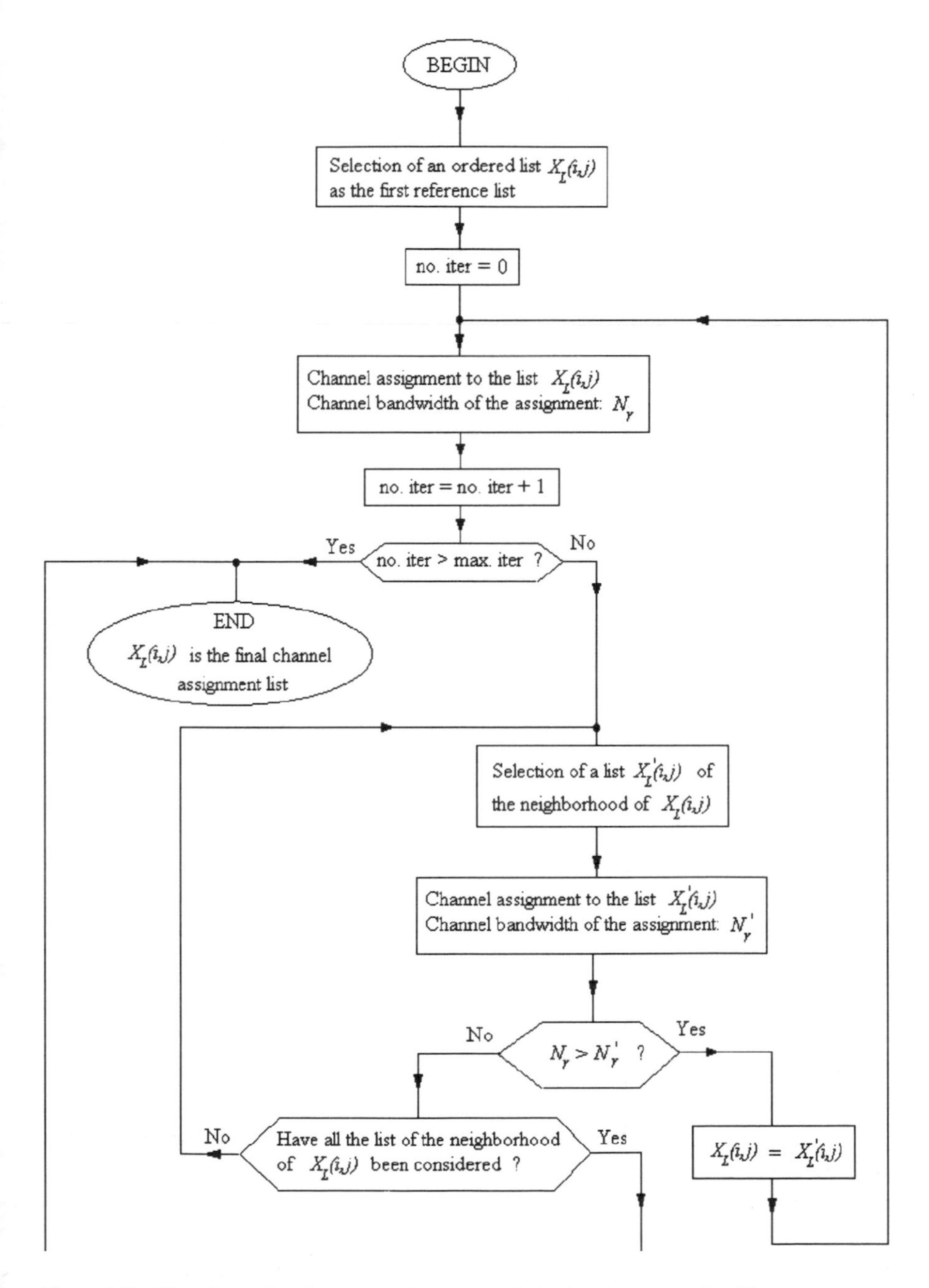

Figure 6.21 Flowchart showing the mains steps of the local-search algorithms.

Table 6.4
Two Neighboring Lists **X** and **X′** and Their Respective Frequency Vectors

l	$X'_L(l,j)$ $X_L(l,j)$; j=1,2	$F(l)$	$X'_L(l,j)$; j=1,2	$F(l)$
1	3, 1	1	1,1	1
2	2, 1	3	2, 1	4
3	1,1	6	3,1	2
4	1,2	10	1,2	7
5	2,2	7	2,2	10
6	3,2	5	3, 2	6
7	1,3	14	1,3	13
8	3, 3	9	3, 3	12

channels by all the BSs of the system. No channel is assigned to a particular microcell, instead all the channels are placed into a pool; in each instant of time a channel is assigned to a particular cell depending on the traffic conditions in the complete system.

The differences among the DCA algorithms are, mainly, in the optimization criterion followed to distribute the channels among the BSs. These differences lie between two extremes.

At one extreme are those algorithms such as the maximum packing (MP) technique [41], which minimizes the number of channels used by the system at all times. This is equivalent to solving a FCA problem each time a new call arrives or finishes. The comment to this technique is that the price to pay for optimizing the channel occupancy is a very high computational cost of reassigning the channels, which might not be affordable for large systems.

In the opposite extreme of the MP algorithms, the channels are allocated to the incoming calls following a more rapid and straightforward procedure: When a new call arrives, we assign it the first available (FA) channel we find that is compatible with all the calls in progress. The FA algorithm is very easy to implement, especially in systems where the channel assignment control is distributed among the microcells. However, the FA solution will not give an optimum channel distribution for future incoming calls. The FA will probably select a channel that will block as incompatible, a large amount of the channels not in use.

Here, we examine a technique that can be considered to be midway between the MP and the FA algorithms. This technique is documented in [39, 40, 42] and can be considered an improvement over the approach described in [13].

In this approach, an initial FCA solution, $Fi(l)$, is taken as a reference pattern in the application of the DCA algorithm. It is assumed that $Fi(l)$

satisfies, in a time average, the traffic demand in the system. In an instant of time, the system is carrying a set of calls. If a new call arrives to the system, we will assign a channel such that (1) it shall be compatible with the channels in use and (2) it shall block the minimum number of possible channels for future incoming calls in accordance with $Fi(l)$. Analogously, if a call finishes, the channel that is released is chosen in such a way that it makes a maximum number of channels available for future incoming calls, also taking $Fi(l)$ as reference.

Let $I(M)$ be the set of microcells that interferes with cell number M. The cell K belongs to $I(M)$ if the term $CM(M, K)$ of the compatibility matrix does not vanish. Let's define $U(M)$ as all the channels in use in microcells M and $I(M)$. The set $N_0(k)$ will be all the channels assigned in the initial FCA to microcell K. Let $A(M)$ indicate all the channels available for new calls in cell M. Set $A(M)$ is formed by all the channels not in use in microcells M and $I(M)$ and which are compatible with channels $U(M)$. The channel to assign to an incoming call in cell M will be chosen among all the channels of $A(M)$, which preferably belong to $N_0(M)$ and blocks the minimum number of channels of $A(K)$ and $N_0(K)$, where K is any cell of $I(M)$. The channel chosen, l, will be the one that minimizes the so-called allocation cost function defined as follows:

$$COA(M, l) = \sum_{K \in I(M)} COAP(M, K, l) \tag{6.10}$$

where $COAP(M, K, l)$ is the cost contribution of microcell K to use channel l in cell M

$$COAP(M, K, l) = \sum_{l+CM(M,K)-1}^{l-CM(M,K)+1} \{v(K, i) + 2[1 - q(K, i)]\} \quad \text{for } M \neq K \tag{6.11a}$$

$$COAP(M, M, l) = q(K, l) + \sum_{l+CM(M,M)-1,\, i \neq l}^{l-CM(M,M)+1} \{v(K, i) + 2[1 - q(K, i)]\} \tag{6.11b}$$

where

$$v(K, i) = \begin{cases} 1 & \text{if } i \in A(K) \\ 0 & \text{elsewhere} \end{cases} \tag{6.12}$$

$$q(K, i) = \begin{cases} 0 & \text{if } i \in N_0(K) \\ 1 & \text{elsewhere} \end{cases} \tag{6.13}$$

Function $v(K, i)$ penalizes with a weight of 1 if the chosen channel l belongs to the channel available for microcell K. Function $q(K, i)$ increases the cost of choosing l if this channel blocks the FCA channels of K, except for the case $q(M, l)$, for which no penalty is added because l belongs to the FCA channels of cell M.

Once a channel has been chosen, all of the sets and parameters that define the system state ($I(M)$, $U(M)$, $A(M)$, etc.) are recomputed for the next event, which can be an incoming call or a released call. Assuming a call is released in microcell M, the DCA algorithm must find the channel to be deallocated among all the channels in use, $U(M)$. We will choose the channel l whose deallocation releases a maximum number of blocked channels. To do that, we choose the channel that minimizes the deallocation cost function defined as follows:

$$DCOA(M, l) = \sum_{K \in I(KM)} DCOAP(M, K, l) \tag{6.14}$$

where $DCOAP(M, K, l)$ is the cost contribution of microcell K when channel l is released in microcell M

$$DCOAP(M, K, l) = \sum_{l+CM(M,K)-1}^{l-CM(M,K)+1} [b(K, i) + 2q(K, i)] \quad \text{for } M \neq K \tag{6.15a}$$

$$DCOAP(M, M, l) = 1 - q(K, l) + \sum_{l+CM(M,M)-1,\, i \neq l}^{l-CM(M,M)+1} [b(K, i) + 2q(K, i)] \tag{6.15b}$$

where

$$b(K, i) = \begin{cases} 0 & \text{if } i \in BK(M, K) \\ 1 & \text{elsewhere} \end{cases} \tag{6.16}$$

where $BK(M, K)$ is the set of channels that are blocked in cell K only by incompatible channels in use in cell M. We note that function $BK(M, K)$ helps to release the maximum number of channels for future incoming calls.

References

[1] Lee, W. C. Y., *Mobile Cellular Telecommunications Systems,* New York: McGraw-Hill, 1990.

[2] Lee, W. C. Y., *Mobile Communications Design Fundamental,* 2nd ed., New York: John Wiley, 1993.

[3] Jabbari, B., "Fixed and Dynamic Channel Assignment," Chap. 21 in *The Mobile Communications Handbook,* J. D. Gibson, Ed., Boca Raton, FL: CRC Press and IEEE Press, 1996.

[4] Katzela, I., and M. Naghhineh, "Channel Assignment Schemes for Cellular Mobile Telecommunication Systems, A Comprehensive Survey," *IEEE Personal Communications,* Vol. 3, No. 3 , June 1996, pp. 10–31.

[5] Frullone, M., G. Riva, P. Grazioso, and G. Falciasecca, "Advanced Planning Criteria for Cellular Systems," *IEEE Personal Communications,* Vol. 3, No. 6, Dec. 1996, pp. 10–15.

[6] Abramson, N., "Multiple Access Techniques for Wireless Networks," *Proc. IEEE,* Vol. 82, Sep. 1994, pp. 1360–1370.

[7] Steele, R., *Mobile Radio Communication,* London: Pentech Press, 1992.

[8] Pichna, R., and Q. Wang, "Power Control," Chap. 23 in *The Mobile Communications Handbook,* J. D. Gibson, Ed., Boca Raton, FL: CRC Press and IEEE Press, 1996.

[9] Yacoub, M. D., "Cell Design Principles," Chap. 19 in *The Mobile Communications Handbook,* J. D. Gibson, Ed., Boca Raton, FL: CRC Press and IEEE Press, 1996.

[10] Steele, R., "Microcellular Radio Communications," Chap. 20 in *The Mobile Communications Handbook,* J. D. Gibson, Ed., Boca Raton, FL: CRC Press and IEEE Press, 1996.

[11] Hernando-Rabanos, J. M., *Comunicaciones Moviles,* Madrid: Editorial Centro de Estudios Ramón Areces, 1997 (in Spanish).

[12] Wong, W., and C. K. Rushforth, "An Adaptive Local-Search Algorithm for the Channel-Assignment Problem (CAP)," *IEEE Trans. on Vehicular Technology,* Vol. 45, No. 3, Aug. 1996, pp. 459–466.

[13] Del Re, E., R. Fantacci, and G. Giambene, "Handover and Dynamic Channel Allocation Techniques in Mobile Cellular Networks," *IEEE Trans. on Vehicular Technology,* Vol. 44, No. 2, May 1995, pp. 229–236.

[14] Everitt, D., "Traffic Engineering of the Radio Interface for Cellular Mobile Network," *Proc. IEEE,* Vol. 82, Sep. 1994, pp. 1371–1382.

[15] Smith, C., and C. Gervelis, *Cellular System. Design & Optimization,* New York: McGraw-Hill, 1996.

[16] El-Dolil, S. A., W. C. Wong, and R. Steele, "Teletraffic Performance of Highway Microcells with Overlay Macrocell," *IEEE J. on Selected Areas in Communications,* Vol. 7, Jan. 1989, pp. 71–78.

[17] Steele, R., and J. E. B. Williams, "Third Generation PCN and the Intelligent Multimode Mobile Portable," *IEE Electronic and Communication Eng. J.,* Mar. 1990, pp. 147–156.

[18] Steele, R., M. Nofal, and S. A. El-Dolil, "An Adaptive Algorithm for Variable Teletraffic Demand in Highway Microcells," *Electronic Lett.,* Vol. 26, No. 14, July 1990, pp. 988–990.

[19] Kahwa, T. J., and N. Georganas, "A Hybrid Channel Assignment Scheme in Large Scale Cellular—Structured Mobile Communication Systems," *IEEE Trans. on Communications,* Vol. 26, 1978, pp. 432–438.

[20] MacDonald, V. H., "Advanced Mobile Phone Service, The Cellular Concept," *Bell Syst. Tech . J.,* Vol. 58, Jan. 1979, pp. 15–41.

[21] Zhang, M., and T. S. Yum, "The Non-Uniform Compact Pattern Allocation Algorithm for Cellular Mobile System," *IEEE Trans. on Vehicular Technology*, Vol. 40, No. 3, 1991, pp. 387–391.

[22] Cox, D. C., and D. O. Reudink, "Dynamic Channel Assignment in High-Capacity Mobile Communications Systems," *Bell System Tech. J.*, Vol. 50, No. 6, July/Aug. 1971, pp. 1833–1857.

[23] Tuttlebee, W. H. W., "Cordless Personal Communications," *IEEE Communication Mag.*, Vol. 82, Sep. 1994, pp. 1360–1370.

[24] Hong, D., and S. S. Rappaport, "Traffic Model and Performance Analysis for Cellular Mobile Radio Telephone Systems with Prioritized and Nonprioritized Handoff Procedures," *IEEE Trans. on Vehicular Technology*, Vol. 35, No. 3, 1986, pp. 77–92.

[25] Oh, S.-H., and D.-W. Tcha, "Prioritized Channel Assignment in a Cellular Radio Network," *IEEE Trans. on Communications*, Vol. 40, No. 7, July 1992, pp. 1259–1269.

[26] Rustako, A. J., N. Amitay, G. J. Owens, and R. S. Roman, "Radio Propagation at Microwave Frequencies for Line-of Sight Microcellular Mobile and Personal Communications," *IEEE Trans. on Vehicular Technology*, Vol. 40, No. 1, Feb. 1991, pp. 203–210.

[27] Tan, S. Y., and H. S. Tan, "A Microcellular Communications Propagation Model Based on the Uniform Theory of Diffraction and Multiple Image Theory," *IEEE Trans. on Antennas and Propagation*, Vol. 44, No. 10, Oct. 1996, pp. 1317–1326.

[28] Cátedra, M. F., J. Pérez, A. González, O. Gutiérrez, and F. Sáez de Adana, "Fast Computer Tool for the Analysis of Propagation in Urban Cells," *Proc. Wireless Communications Conference*, Boulder, CO, Aug. 11–13, 1997, pp. 240–245.

[29] Cátedra, M. F., J. Pérez, F. Saez de Adana, and O. Gutierrez, "Efficient Ray-Tracing Techniques for 3D Analysis of Propagation in Mobile Communications. Application to Picocell and Microcell Scenarios," *IEEE Antennas and Propagation Mag.*, Vol. 40, No. 2, April 1998, pp. 15–28.

[30] Kouyoumjiam, R. G., "A Uniform Geometrical Theory of Diffraction for an Edge in a Perfectly Conducting Surface," *Proc. IEEE*, Aug. 1974, pp. 1448–1461.

[31] Kanatas, A. G., I. D. Kountoris, G. B. Kostaras, and Ph. Constantinou, "A UTD Propagation Model in Urban Microcellular Environments," *IEEE Trans. on Vehicular Technology*, Vol. 46, No. 1, February 1997, pp. 185–193.

[32] Maric, S., and I. Seskar, "Microcell Planning and Channel Allocation for Manhattan Street Environments," *Proc. 1st ICUP*, Dallas, TX, Sep. 1992.

[33] Goldsmith, A. J., and L. J. Greenstein, "A Measurement-Based Model for Predicting Coverage Areas of Urban Microcells," *IEEE J. of Selected Areas in Communications*, Vol. 11, Sep. 1993, pp. 1013–1023.

[34] Clark, M. V., V. Erceg, and L. Greenstein, "Reuse Efficiency in Urban Microcellular Networks," *IEEE Trans. on Vehicular Technology*, Vol. 46, May 1997, pp. 279–288.

[35] Funabiki, N., and Y. Takefuji, "A Neural Network Parallel Algorithm for Channel Assignment Problems in Cellular Radio Networks," *IEEE Trans. on Vehicular Technology*, Vol. 41, No. 4, Nov. 1992, pp. 430–437.

[36] Kim, S., and S.-L. Kim, "A Two-Phase Algorithm for Frequency Assignment in Cellular Mobile Systems," *IEEE Trans. on Vehicular Technology*, Vol. 43, No. 3, Aug. 1994, pp. 542–548.

[37] Sivarajan, K. S., R. J. McEliece, and J.W. Ketchum, "Channel Assignment in Cellular Radio," *Proc. 39th IEEE Vehicular Technology Conference,* May 1989, pp. 846–850.

[38] Zoellner, J. A., and C. A. Beall, "A Breakthrough in Spectrum Conserving Frequency Assignment Technology," *IEEE Trans. Electromagn. Compat.,* Vol. 9, Aug. 1977, pp. 313–319.

[39] Gonzalez, I., "Channel Assignments Techniques for Cellular Systems," Master's Thesis, University of Cantabria, Oct. 1997.

[40] Cantalapiedra, J., "Computer Tools Integration for the Analysis and Design of Cellular Systems," Master's Thesis, University of Cantabria, Oct. 1997.

[41] Raymond, P. A., "Performance Analysis of Cellular Networks," *IEEE Trans. on Communications,* Vol. 39, No. 12, Dec. 1991, pp. 1787–1793.

[42] Cantalapiedra, J., I. Gonzalez, and M. F. Cátedra, "Dynamic Channel Allocation and Performance Analysis in Mobile Cellular Systems," Internal Report, University of Alcalá.

[43] Jakes, W. C., *Microwave Mobile Communications,* New York: John Wiley, 1974.

Appendix 6A: Demand Vector Computation

The total area of the wireless system is split into a grid of small rectangles, $S_\Delta(i,j)$, of an area $\Delta x \Delta y$. The center of rectangle $S_\Delta(i,j)$ is given by

$$r_{ij} = i\Delta\hat{x} + j\Delta\hat{y} \tag{6A.1}$$

Points r_{ij} will be referred to as *sampling points.* The BS number M is located at point r_M and radiates a maximum power of V_M(dBm). The power level at r_{ij} due to the BS number M can be computed using:

$$W_M(i,j) = V_M - L_M(i,j) \qquad \text{(dBm)} \tag{6A.2}$$

where $L_M(i,j)$ includes all the propagation losses and the antenna gains.

The area S_M, belonging to microcell M, is formed by all the rectangles in which the power received from BS M is greater than the power received from any other BS. Mathematically, we have

$$S_\Delta(i,j) \in S_M \text{ if } W_M(i,j) \geq W_N(i,j) \quad \text{for } N = 1, 2, \ldots, N_T; \quad N \neq M \tag{6A.3}$$

where N_T is the total number of BS. S_M can also be specified with the help of functions $SD_M(i,j)$, where

$$SD_M(i,j) = \begin{cases} 1 & \text{if } W_M(i,j) \geq W_N(i,j) \quad \text{for } N = 1, 2, \ldots, N_T; \quad N \neq M \\ 0 & \text{elsewhere} \end{cases} \tag{6A.4}$$

The traffic (in Erlangs) required by the potential subscribers in S_M will be

$$T(M) = \Delta x \Delta y \sum_{i=i\min}^{i\max} \sum_{j=j\min}^{j\max} SD_M(i,j)\, t(i,j) \tag{6A.5}$$

where $t(i,j)$ is the estimated traffic at point r_{ij} per surface unit at point r_{ij}. Two kinds of traffic should be considered: the traffic for new call origination and traffic for the handoffs. The traffic density for new calls can be considered to be uniformly distributed over the microcell areas. The traffic for handoffs depends on the mobility of the subscribers, the duration of the message, and the common boundaries between microcells [24, 43]. A first estimation for both traffic distributions is to consider their traffic densities, $t(i,j)$, as being a constant on the entire system surface.

The demand vector, $V(M)$, can be obtained using formula B of Erlang, assuming that the new calls are dropped if all the channels of BS M are busy. For a specified blocking probability P_b, vectors $V(M)$ and $T(M)$ are related by

$$P_b = \frac{\dfrac{(T(M))^{V(M)}}{(V(M))!}}{\displaystyle\sum_{i=0}^{V(M)} \frac{(T(M))^i}{i!}} \tag{6A.6}$$

In this last equation, vectors $V(M)$ and $T(M)$ are assumed to be rounded up to their closest higher integer values.

Appendix 6B: Interference Matrix Computation

The compatibility matrix, $CM(M, N)$, will be obtained from the interference matrix $IM(M, N)$. The entry $IM(M, N)$ is the C/I value so that, at least, in 90% of the points of S_M, the ratio between the power received from a desired channel of BS M and the interference power of an undesired channel of BS N is equal to or greater than that C/I value (90% criterion). Mathematically, we have

$$F_{MN}[CI_{MN} \leq IM(N, M)] = 0.1 \tag{6B.1}$$

where F_{MN} is one cumulative probability function that we can associate with the C/I ratio:

$$CI_{MN}(i,j) = W_M(i,j) - W_N(i,j) \text{ (dB)} \tag{6B.2}$$

defined over the sampling points (i,j), which belong to cell M.

The computation of $IM(M,N)$ changes if a power control is operative. We will see how to compute the interference matrix with and without power control.

6B.1 Computation of *MI*(*M,N*) Without Power Control

6B.1.1 Diagonal Terms *IM*(*M,M*)

In this case the desired and the undesired signals have the same power level at all of the sampling points. Therefore, all the sampling points have the same C/I:

$$CI_{MM}(i,j) = IM(M,M) = 0 \text{ dB} \tag{6B.3}$$

6B.1.2 Arbitrary Terms *IM*(*M,N*), $M \neq N$

Now, the value of $IM(M,N)$ is found by applying (6B.1) and (6B.2).

6B.2 Computation of *IM*(*M,N*) With Power Control

6B.2.1 Diagonal Terms *IM*(*M,M*)

Thanks to the power control, the transmitter power can be reduced and therefore the C/I ratio in the sampling point (i,j) can be different from 0 dB. We will call $\Delta M(i,j)$ and $\Delta' M(i',j')$ the reductions in the transmitted power for MSs located at points (i,j) and (i',j'), respectively. The ratio $CI_{MM'}(i,j)$, between the desired power for an MS at (i,j) and the undesired power this MS is receiving due to a channel used by an MS located at (i',j') is given by

$$\begin{aligned} CI_{MM'}(i,j) &= [W_M(i,j) - \Delta M(i,j)] - [W_{M'}(i,j) - \Delta' M(i',j')] \\ &= \Delta' M(i',j') - \Delta M(i,j) \end{aligned} \tag{6B.4}$$

where ΔM and $\Delta' M$ are the decreases in the transmitted power of the desired and undesired signals, respectively, due to the power control. Again, we can assume a random variable, $CI_{MM'}$, linked to the function defined in (6B.4).

The cumulative probability function for $CI_{MM'}$ can be obtained from the cumulative density functions $F_{\Delta M}(\Delta M)$ and $F_{\Delta' M}(\Delta' M)$, which we can form considering ΔM and $\Delta' M$ as random variables:

$$F_{MM'}(CI_{MM}) = \int_0^{\Delta' M\max} F_{\Delta M}(\Delta' M - CI_{MM}) f_{\Delta' M}(\Delta' M) d\Delta' M \quad (6B.5)$$

where $\Delta' M\max$ is the maximum value allowable for $\Delta' M$ and $f_{\Delta' M}(\Delta' M)$ is the probability density function of $\Delta' M$. Applying the 90% criterion, the diagonal terms of matrix IM are obtained:

$$F_{MM'}[CI_{MM'} \leq IM(M, M)] = 0.1 \quad (6B.6)$$

6B.2.2 Arbitrary Term *IM*(*M*,*N*), *M* ≠ *N* Computation With Power Control

First, we define the random variable $CI_{MM'}$, which gives the C/I ratio considering that no power control is applied in microcell N:

$$CI_{MN'} = CI_{MN} - \Delta M \quad (6B.7)$$

where CI_{MN} is the random variable associated with function $CI_{MN}(i,j)$ defined in (6B.2) and ΔM is as defined in the previous section. The cumulative probability function for $CI_{MM'}$ is defined by

$$F_{MN'}(CI_{MN'}) = \int_0^{\Delta M\max} F_{MN}(CI_{MM'} + \Delta M) f_{\Delta M}(\Delta M) d\Delta M \quad (6B.8)$$

Then, the random variable $CI_{MN''}$, which represents the C/I at microcell M when the power control is activated in both microcells M and N, can be defined as

$$CI_{MN''} = CI_{MM'} + \Delta' M \quad (6B.9)$$

The cumulative probability function for $CI_{MM''}$ is now given by

$$F_{MN''}(CI_{MN''}) = \int_0^{\Delta' M\max} F_{MN'}(CI_{MM''} - \Delta' M) f_{\Delta' M}(\Delta' M) d\Delta' M \quad (6B.10)$$

By applying the 90% criterion, the off-diagonal terms of matrix IM are obtained solving

$$F_{MN''}[CI_{MN''} \leq IM(M, N)] = 0.1 \qquad (6B.11)$$

Appendix 6C: Compatibility Matrix Computation

From the interference matrix and from the particular parameters of the communication system (kind of modulation, multiple access, etc.), it is easy to compute the compatibility matrix, $CM(M, N)$.

One of the significant datum of the system is the minimum value of the carrier to interference ratio, $(C/I)_0$, which guarantees acceptable quality in the desired signal. If the term $IM(M, N)$ of the interference matrix is greater than $(C/I)_0$, then microcells M and N are completely compatible; that means that the same channel can be reused in both cells. The term $CM(M, N)$ is made equal to zero for all the pairs of microcells, which allows the reuse of channels:

$$CM(M, N) = 0 \qquad \text{if } IM(M, N) \geq (C/I)_o \qquad (6C.1)$$

If the term $IM(M, N)$ is less than $(C/I)_o$, then we would have cochannel interference if the same channel is used in microcells M and N. In this case, we can even have, in a FDMA system, adjacent-channel interference if channels close in frequency are used. The interfering power level is mitigated by the additional losses of the receiver front-end filter. For two channels a and b, separated in frequency by $n\Delta f$, where Δf is the bandwidth of the channels, the additional losses are given by

$$L(n) = \frac{K}{0.3}\log_{10}\left(\frac{n\Delta f}{\Delta f}\right) = \frac{K}{0.3}\log_{10} n \qquad (6C.2)$$

where K is the slope of the front-end filter in decibels per octave and n is the number of channels between channels a and b.

The term $CM(M, N)$ is now found from

$$IM(M, N) + L(n_o) = (C/I)_o \qquad (6C.3)$$

$$CM(M, N) = n_o \qquad (6C.4)$$

Sometimes the compatibility matrix is approximated to the nearest higher integer

$$CM(M, N) = \mathrm{INT}(n_o) \tag{6C.5}$$

Usually, this matrix is symmetrical.

About the Authors

Manuel F. Cátedra received his M.S. and Ph.D. degrees in telecommunications engineering from the Polytechnic University of Madrid (UPM) in 1977 and 1982, respectively. From 1976 to 1989 he was with the Radiocommunication and Signal Processing Department of UPM, teaching and doing research. He has been a professor at the University of Cantabria from 1989 to 1998. Currently he is a professor at the University of Alcalá, in Madrid, Spain.

He has worked on approximately 25 research projects solving problems of electromagnetic compatibility in radio and telecommunication equipment, antennas, microwave components and radar cross sections, and mobile communications. He has developed and applied CAD tools for radio equipment systems on naval ships, aircraft, helicopters, and satellites; for contractors from Spanish and European institutions such as CASA, ALCATEL, DASA, SAAB, INTA, BAZAN, the Spanish Defense Department, and the French company, MATRA. He is currently working on a project for Telefonica (the largest Spanish telecommunication company) to develop computer tools for propagation analyses in microcells and indoor cells.

He has directed approximately 12 Ph.D. dissertations, published approximately 35 papers (published in *IEEE, Electronic Letters,* and *Applied Computational Electromagnetics Society Journal*), written a book, *The CG-FFT Method. Application of Signal Processing Techniques to Electromagnetics* published by Artech House in 1995, as well as contributed approximately 10 chapters in different books, given short courses, and has given approximately 100 presentations at international symposia.

Jesús Pérez-Arriaga received his M.S. and Ph. D degrees in electronic engineering from the University of Cantabria, Spain. Since 1993 he has been an assistant professor at the University of Cantabria. In 1989 he was with the Radiocommunication and Signal Processing Department of the Polytechnic University of Madrid as a research assistant. From 1990 to 1992 he was with the Electronics Department of the University of Cantabria as a research assistant. In 1993 he became an assistant professor in the Electronics Department of the University of Cantabria. Since October 1998 he has been an Assistant Professor at the University of Acalá in Madrid, Spain. He has participated in 15 research projects related to RCS computation with Spanish and European companies, performed analysis of on-board antennas and radio propagation in mobile communications. He is the author of 10 papers and of more than 20 conference contributions at international symposia. His research interests include high-frequency methods in electromagnetic radiation and scattering, and mobile communications.

Index

Recent Titles in the Artech House Mobile Communications Series

John Walker, Series Editor

Advances in Mobile Information Systems, John Walker, editor

An Introduction to GSM, Siegmund M. Redl, Matthias K. Weber, Malcolm W. Oliphant

CDMA for Wireless Personal Communications, Ramjee Prasad

CDMA RF System Engineering, Samuel C. Yang

CDMA Systems Engineering Handbook, Jhong S. Lee and Leonard E. Miller

Cell Planning for Wireless Communications, Manuel F. Cátedra and Jesús Pérez-Arriaga

Cellular Communications: Worldwide Market Development, Garry A. Garrard

Cellular Mobile Systems Engineering, Saleh Faruque

The Complete Wireless Communications Professional: A Guide for Engineers and Managers, William Webb

GSM and Personal Communications Handbook, Siegmund M. Redl, Matthias K. Weber, Malcolm W. Oliphant

GSM Networks: Protocols, Terminology, and Implementation, Gunnar Heine

GSM System Engineering, Asha Mehrotra

Handbook of Land-Mobile Radio System Coverage, Garry C. Hess

Introduction to Radio Propagation for Fixed and Mobile Communications, John Doble

Introduction to Wireless Local Loop, William Webb

IS-136 TDMA Technology, Economics, and Services, Lawrence Harte, Adrian Smith, Charles A. Jacobs

Mobile Communications in the U.S. and Europe: Regulation, Technology, and Markets, Michael Paetsch

Mobile Data Communications Systems, Peter Wong, David Britland

Mobile Telecommunications: Standards, Regulation, and Applications, Rudi Bekkers and Jan Smits

Personal Wireless Communication With DECT and PWT, John Phillips, Gerard Mac Namee

RDS: The Radio Data System, Dietmar Kopitz, Bev Marks

RF and Microwave Circuit Design for Wireless Communications, Lawrence E. Larson, editor

Spread Spectrum CDMA Systems for Wireless Communications, Savo G. Glisic, Branka Vucetic

Understanding Cellular Radio, William Webb

Understanding Digital PCS: The TDMA Standard, Cameron Kelly Coursey

Understanding GPS: Principles and Applications, Elliott D. Kaplan, editor

Universal Wireless Personal Communications, Ramjee Prasad

Wideband CDMA for Third Generation Mobile Communications, Tero Ojanperä, Ramjee Prasad

Wireless Communications in Developing Countries: Cellular and Satellite Systems, Rachael E. Schwartz

For further information on these and other Artech House titles, including previously considered out-of-print books now available through our In-Print-Forever® (IPF®) program, contact:

Artech House
685 Canton Street
Norwood, MA 02062
Phone: 781-769-9750
Fax: 781-769-6334
e-mail: artech@artech-house.com

Artech House
46 Gillingham Street
London SW1V 1AH UK
Phone: +44 (0)171-973-8077
Fax: +44 (0)171-630-0166
e-mail: artech-uk@artech-house.com

Find us on the World Wide Web at:
www.artechhouse.com